THE SCIENCE OF LAST THINGS

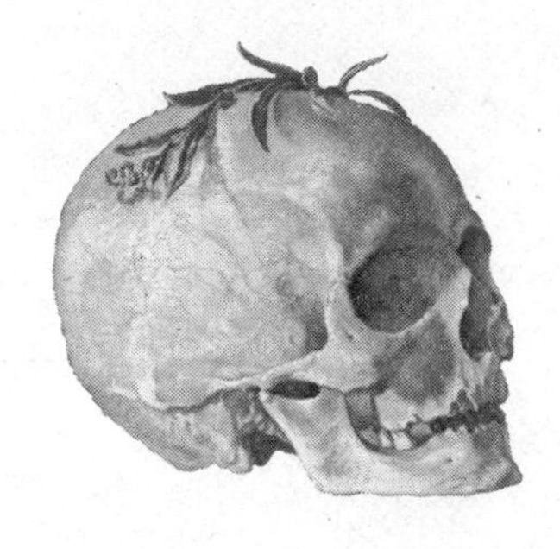

THE SCIENCE OF LAST THINGS

essays on deep time and the boundaries of the self

Ellen Wayland-Smith

MILKWEED EDITIONS

Published 2024 by Milkweed Editions
Printed in Canada
Cover design by Mary Austin Speaker
Author photo by Sarah Wayland-Smith
24 25 26 27 28 5 4 3 2 1
First Edition

Library of Congress Cataloging-in-Publication Data

Names: Wayland-Smith, Ellen, 1966- author.
Title: The science of last things : essays on deep time and the boundaries of the self / Ellen Wayland-Smith.
Description: Minneapolis, Minnesota : Milkweed Editions, 2024. | Includes bibliographical references. | Summary: "This luminous collection of essays explores life's liminal states-such as losing a parent, birthing another body, experiencing a nervous breakdown and faces the unescapable reality of our bodily impermanence" -- Provided by publisher.
Identifiers: LCCN 2024002122 (print) | LCCN 2024002123 (ebook) | ISBN 9781639550968 (paperback) | ISBN 9781639550975 (ebook)
Subjects: LCSH: American essays--21st century. | LCGFT: Essays.
Classification: LCC PS3623.A956 S35 2024 (print) | LCC PS3623.A956 (ebook) | DDC 814/.6--dc23/eng/20240805
LC record available at https://lccn.loc.gov/2024002122
LC ebook record available at https://lccn.loc.gov/2024002123

Milkweed Editions is committed to ecological stewardship. We strive to align our book production practices with this principle, and to reduce the impact of our operations in the environment. We are a member of the Green Press Initiative, a nonprofit coalition of publishers, manufacturers, and authors working to protect the world's endangered forests and conserve natural resources. *The Science of Last Things* was printed on acid-free 100% postconsumer-waste paper by Friesens Corporation.

for my mother, Kate, and my sister, Sarah,
with all my love

CONTENTS

Preface

HOW TO LIVE IN DEEP TIME

In my backyard, I have a makeshift pond: a galvanized steel tub about two feet in diameter, outfitted with an electric filter, a heating coil, algae-skimmed driftwood, and a paving stone propped up on two overturned clay flowerpots. It is a summer day, and from where I sit, reading in the shade, I hear the filter bubble softly as the sun crests the roof of the house. Soon, I hear a flipper-y noise—a muffled knock and splash of water against the tub's sides. I put down my book.

Sweeney, our three-year-old red-eared slider turtle, hoists himself—neck straining, claws scraping—up out of the water and onto the sun-soaked paving stone for his afternoon bask.

Perched motionless, head tilted toward the sun's warming rays, Sweeney appears either magnificent or absurd, depending on my mood. His neck is swamp green, thick as a celery stick and soft as a lamb's ear. His domed shell is as rough and ridged as damp tree bark. When I approach him—slowly, slowly, lest he startle—he fixes me with eyes like two tiny wells of liquid onyx rimmed in blazing lime green. He blinks. I hold out a blueberry, and he cracks open his beak to reveal a triangle of tongue, the softest bubblegum pink.

Many traditional cultures have myths about how the turtle got its shell. In Aesop's fable, the turtle, loath to leave the comfort of her home, declines an invitation to Zeus's wedding celebration. For this breach of etiquette, Zeus condemns her to wander the world lugging her home on her back in perpetuity. Other versions of the myth are less punishing. For the Anishinaabe, the turtle receives a beautifully painted stone shell in return for showing the god Nanabozho the best fishing spots in the river.

But evolutionary biology's origin story is no less fantastic. The turtle's precise phylogenetic origins are up for debate, but at some point during the Triassic, as long as 240 million years ago, what one biologist hypothesizes was a "stout lizard" began to convert the bones of its sternum into a protective body plate. Such an armored underbelly granted the creature an evolutionary edge; over time the prototurtle's ribs, too, flattened and fused, turned spatulate, knit together into an elegantly curved carapace. Sweeney and his ilk are what biologists call a "highly conserved" species, an evolutionary success story. Indeed, Sweeney survived multiple mass extinction events, including the asteroid hit that wiped out the dinosaurs, scrambling up onto a triumphal rock as his reptilian cousins perished all around him.

As far as pets go, Sweeney is both more work (maintaining his outdoor aquatic environment is exhausting) and less rewarding (no cuddles) than your average mammalian pet, a dog or a cat. Yet to keep company with a turtle, archaic reminder of life's murkier, rockier, hard-scrabble origins, has been instructive. In her poem "Landcrab I," Margaret Atwood addresses a hermit crab on a beach as secret sharer of her being. The scowling creature, Atwood muses, has "nothing but contempt" for mammals, "with their lobes and tubers/scruples and warm milk." And yet in this "stunted child," born of "dragon [. . .] teeth," she nonetheless recognizes "a piece of what we are."

As befits the quaint domesticity of the Victorian era, Charles Darwin imagined the origins of life on earth as a cozy affair, all lobes and warm milk. In an 1871 letter to Joseph Hooker, Darwin conjured a scene of antediluvian innocence, "some warm little pond" where, out of a rich stew of "ammonia and phosphoric salts, light, heat, electricity," a protein chain quivered and quickened. Yet over the past few decades, research on deep-sea thermal vents has led scientists to speculate that it was not in the rustic country tarn of Darwin's imagination but rather in the Archaean depths of a lightless ocean trench, on the mineral lip of a hydrothermal vent, that life first took shape more than 3.8 billion years ago: more dragon's tooth than amniotic sac. The solid surfaces of oceanic mineral and rock provided a base onto which a protomembrane could perch long enough to self-organize into a fragile bubble that would stake its claim against the surrounding chaos, an islet of organization in the flux of the new earth.

The image of this original bubble, solitary pocket of inwardness winking in the deep, transfixes me and is my preferred creation myth. It reminds me that any single biological life-form is defined not by the elements that make it up, but by how the arrangement of those elements inside a membrane differs from their arrangement outside of it and how successful that life-form is at maintaining this difference across time. I am a bundle of hydrogen and oxygen and carbon and nitrogen and calcium and phosphorous atoms that differs from the surrounding sea of those same primary elements only by dint of organization and complexity, not substance; and, in fact, over the course of my life I will swap out every atom of my original molecular makeup for atoms pulled in from outside myself, building myself from scratch like Theseus's ship. A living organism is not so much a substance as a concentration gradient; not so much a thing as a proportional relationship maintained through the doggedness of habit.

Living organisms are hardwired to spend every ounce of energy they have maintaining this gradient. The instant the organizational difference separating my little bubble of molecules from the surrounding sea of molecules collapses, I cease to exist: my elements wash, once again, back into the cosmic elemental broth that is (on the largest scale) the universe expanding toward thermodynamic equilibrium. My own little parcel of organized stuff, digging in like a barnacle on a sea rock—down-river descendent of the original bubble trembling on a vent-seam in the mineral ocean bed—succumbs to the salt tide of disorganization.

The mystery of the self, then, is that every atom of our being is on loan from the universe, at every instant, from time immemorial. This is a fact of which very few of us are ever cognizant in any consistent way. Humans don't much like to acknowledge this loan, most likely because our brains work at time scales supremely unsuited to appreciate the nearly four-billion-year-long game of biological life on earth. We are a prideful species, hesitant to claim kinship with the primeval bubbling ooze that started the whole ball rolling.

All these thoughts are swirling round in my head when suddenly, out of the literal blue, a giant jay comes screeching and thrashing through the pepper tree to peck on seeds at the feeder. Sweeney scuttles back into the water with a soft splash. He returns to the algal murk: drifting soundlessly, poking his nose among driftwood, reprising his mysterious underwater existence. Jay, turtle, tree, me: we are all cut from the same cloth. I return to my book.

To live with one eye fixed on geochemical and biological "deep time" is to practice radical humility, in the most literal sense of the world: *humilitas,* from *humus*, the Latin word for "earth." To be low to the

ground, on a par with the earth's crust, subject to the same cyclical processes of roiling, cooling, ribboning, crumbling, rise and fall and rise again. Geologists become so habituated to this way of counting time that they end up as philosophers. In an interview with John McPhee, geologist David Brower gave a sense of how the study of rocks had shaped his life. "You begin tuning your mind to a time scale that is the planet's time scale," he observed. "For me, it is almost unconscious now and is a kind of companionship with the earth."

These essays are my attempt to follow the seams of deep time as, here and there, they have surfaced along the path of my life. They are my attempt to listen for something beyond the narrow pulse and tick of human clocks: to keep companionship with the earth.

THE SCIENCE OF LAST THINGS

GRAVITY

One morning I was sleeping on the sofa in my parents' apartment when I was woken by the sound of my father dying in the next room.

At first I couldn't tell what the sound was, or even locate where it was coming from. It was a raspy, scraping sound, like metal being pulled through tightly packed glass. Then it shifted: like someone breathing in a viscous liquid in greedy gulps, aspirating yogurt. When I realized the noises were coming from my father's throat, I froze. According to the hospice manual I had scanned the night before, this was the proverbial "death rattle": the sound produced by pooling secretions in the throat as a body's swallow reflex weakens.

We had had almost no time to prepare. A mere ten days earlier, my father had gone in to his doctor's office to pick up the results of a routine scan, which turned out not to be routine at all: stage four pancreatic cancer. His physician, an old family friend, teared up when he delivered the news. "It is very difficult for me to say this to you," he'd begun gingerly. "Not as difficult as it is for me to hear it," my father responded.

He was eighty-one but looked much younger: six-foot-two, straight as a poker, salt-and-pepper hair and beard. After a bout with polio when he was fourteen, he'd never been sick a day in his life. We thought he'd live forever.

My father's fall into death was precipitous. The cancer was inoperable, and he declined further treatment. His doctors estimated he had two months to live. Yet only three days after his diagnosis, he began to act oddly disoriented, forgetting where he was. He had trouble buttoning his shirt, tried to eat a bowl of soup by dipping the handle end of the spoon into it. We attributed these behavioral quirks to anxiety. They turned out to be the result, instead, of small strokes—blood clots thrown off by the tumor. By day five he was home and set up in a hospital bed, with a PICC line, catheter, and morphine pump. He slipped in and out of consciousness. The last time I saw him awake, I asked if he'd like me to read him some Jorge Luis Borges, his favorite writer. "That would be nice," he'd said with a smile. "I love you so much." I squeezed his hand and went to get the book. By the time I came back with it, he was asleep again.

In death, the body is baroque in its flows and suppurations. It contorts and contracts until, finally, it returns to the clay from which it was pulled. It betrays our most earnest attempts to contain it, tame it, make it presentable. My father's body oozed: thick urine, yellow shading into crimson along the foggy catheter. Sputum, blood, saliva mixed, brought up from his lungs like sea foam, spotting his beard and chest with frothy pale green patches the color of lettuce. His eyes secreted tears and mucus in his sleep, an opaque glue that stretched and gapped like tiny spiderwebs across his pupils when he fluttered his eyelids. His arms had bleeds beneath the skin where needles poked, deep branching blues and purples the shade of concord grapes.

The morning I woke to the sound of my father's ragged last breaths, I stumbled down the hallway and into his room. I stared at his laboring chest in disbelief. But by that point, I was used to the dying body's betrayals, its refusals to be decent as it made the harrowing passage from person to thing. The death rattle held no surprises for me. His breaths grew shorter, more chopped, more erratic. When they stopped, a quality of stillness unlike anything I'd ever known flooded the room. I threw myself across his body and wailed in a voice I didn't recognize as my own.

Since my father's death, I can't stop thinking about descents and falls. Tumbles, gaffes, face-plants, sprawls. Once when I was sprinting nimbly up the three steps of my office building to get the door a colleague was kindly propping open for me, my foot slipped and I landed on my knees, scalding coffee streaming down my arms. I gazed up at him from this posture of abject genuflection and felt sorrier for him than for myself at our mutual discomfiture: I started giggling, and I am sure he had to bite his lip to keep from doing the same.

Charles Baudelaire analyzed the peculiar mirth humans feel in the face of a fellow creature falling, tracing the phenomenon of laughter back to Adam and Eve's fall. "[Laughter] is one of the clearest marks of Satan in man, one of the innumerable seeds contained in the original symbolic apple," he suggests. Pride goes before a fall, and laughter is a natural response to the spectacle of puffed-up human dignity toppled from its throne. First you have a grip on your body; then, without warning, gravity has you in its grip: you are, after all, only a thing.

Like Adam and Eve, Lucifer, too, took a spectacular tumble, from the light-spangled dome of heaven to the brimstone of hell. "How art thou fallen from heaven, O Lucifer, son of the morning! [. . .] For thou hast said in thine heart, I will ascend into heaven, I will exalt my throne above the stars of God." God sends Lucifer reeling for his act of prideful rebellion. Readers complained of John Milton's *Paradise Lost* that he made the character of Satan too sympathetic. But what else would Satan be? Part of being a creature means that, from the get-go, you throw every ounce of freedom you have toward resisting your createdness, permanent wound to the ego. "Better to reign in hell, than serve in heaven," the rebel angel consoles himself.

One of the last times I saw my father, we were at our summer cottage on Lake Ontario. It was in August, two months before we would even know about his cancer. The sky was a brilliant, glassy blue; we were all healthy, or at least so it seemed, and happy, warmed inside and out by the late-summer sun.

Usually the lake is a flat expanse of green, but on certain windy days the water whips up into Atlantic-style waves, cresting white. It was one of those days, and my daughters and nephew were splashing in the water when my five-year-old niece asked if I could take her in. The beach down to the lake is rock strewn and, although the stones are rounded by erosion, walking over them involves equal parts balance and tolerance for pain, if one doesn't have beach shoes (which we never did). The feat gets even trickier once you reach the water's edge: the rocks are wet and skimmed with algae, and to reach the sand bar you have to traverse a perilous five-foot stretch where a misplaced foot can send you sprawling. The kids were calling for Grandpa to get wet, too, so I hoisted my niece up onto my hip, and my father and I headed out to join them. We were shin-deep in water when a large wave came washing in, and we both went down.

My hands were occupied securing my niece upright beside me against the oncoming waves: one big wave and her small frame would be dashed against the stones. Every time I tried to pivot myself to standing, I was smacked flat on my spine. My father, too, seemed unable to gain purchase on the slippery rocks—couldn't flip to crawling position in order to right himself.

And Baudelaire was right: there was something hilarious about our plight. The two of us, my father and I, scrabbling around helplessly on our backs in the water like pinned insects, like babies on a changing table, like overturned turtles. But it was unsettling at the same time. How would we ever stand up? The scene struck me as a parable, maybe a riddle—though its hidden meaning remained unclear. "What has four legs in the morning, two in the afternoon, and three in the evening?" the Sphinx asks, and Oedipus solves the riddle: man. But we had spun the oracle's picture on its head, one standing baby and two sprawled adults.

We eventually managed to right ourselves. Released from my prone clutches and set safely up on the shore, the five-year-old toddled off in her tangerine bathing suit. My father's white T-shirt, soaked now, clung to his chest, to his sun-browned arms, like a baptismal robe. Or a shroud.

Falling is perhaps practice for death, when the essential imposture of self-sufficiency that we struggle to maintain throughout life is unveiled once and for all. The Jewish tradition of the kippah or yarmulke acknowledges this truth in its own way. *Kippah* in Hebrew means "dome"; *yarmulke* in Aramaic, "in awe of the King." Don't ever forget who reigns in the light-filled dome above your head. You don't belong to yourself. Remember: you belong to God.

Once when I was a graduate student in Paris, I took the train to the Basilica of Saint-Denis on the northern edge of the city. It is a gothic abbey built on the site where the relics of third-century martyr Saint Denis was buried. For centuries it was the burial place for France's kings and queens, from the seventh century Dagobert through the posthumous statues erected in honor of the unfortunate Louis XVI and Marie Antoinette. But most stunning are the funeral monuments of Louis XII and Queen Anne de Bretagne.

They partake of a style of funerary sculpture popular in the late Middle Ages called the cadaver tomb, or *transi*, portraying the body in its transition through death and decay. These are often double-decker tombs. On top is a statue of the deceased as he or she was in life, kneeling in prayer: composed, beautiful, the very picture of decorous containment. Below, in contrast, is figured the naked body in its death throes. Queen Anne's head is thrown back, her chest thrust up from the pallet, body contorted as though she is struggling to draw breath. Her hair is matted, wild, sticking to the skull with sweat. Louis beside her appears already a death's head, the skin on his face retreating to reveal a grimace of teeth. The soft flesh of his eyes is sinking, sucked toward the knobby bone of the sockets.

Today, almost everything coaxes Americans to forget this side of death. We do our best to move through life by doubling ourselves over, to seal up the flap of being whose rough and bloody edges remind us of our creatureliness. We improvise a posture of containment.

In our rejection of suffering, we do our best to make death disappear. Already in 1885, in *The Adventures of Huckleberry Finn,* Mark Twain could write about the peculiar slickness of the undertaker, whose job is to handle death with kid gloves, to usher its object out of the midst of the living as quietly and discreetly as possible. Huck attends the funeral of Colonel Grangerford, where the undertaker sets up the coffin in the middle of the parlor with chairs fanned out around it. Once the parlor is filled with people, the undertaker "slid around in his black gloves with his softy soothering ways, putting on the last touches, and getting people and things all ship-shape and comfortable, and making no more sound than a cat." In absolute silence, the undertaker orchestrates the viewing: moving people softly this way and that, making room for the late-comers, opening up passageways to ease the mourners' approach to the coffin. He was "the softest, glidingest, stealthiest man I ever see," Huck observes with an admiring whistle.

When my father died, the men from the funeral home down the street had already been alerted we would require their services sometime soon. My uncle called them and within an hour the funeral parlor director and his son appeared in my father's bedroom. They, too, were softy and soothering, responsible primarily for making sure we—the living—were all "ship-shape and comfortable." Stealth was key. Hands clasped respectfully in front of them, heads bowed and studiously avoiding eye contact, they wheeled in a gurney with a burgundy tarp-like wrap opened up on top to receive my father's body. Unlike the theatrical, inky mourning of the Victorians, keening its grief in sable silk and velvet, we wrap our dead in a shade of washed-out cabernet—the institutional dusty rose of waiting rooms, whether hotel lobby or doctor's office—the unobtrusive noncolor of polite bureaucracy.

For most of my life I basically ignored my body. Deal with your own body as discreetly as possible when you can, and when you can't, hand it over to the experts for as short a time as possible, was my pragmatic philosophy. And so, when it came time to choose a birthing plan for my first baby, the prospect of natural childbirth held no charms for me. I chose the epidural, and for my first child, it worked. With my legs hanging like wet noodles in the stirrups, and with the doctor's struggles at my nether regions blocked from view by a clean white sheet spread across my knees, I gave a few half-hearted pushes and ended up a few minutes later with a plump, rosy-cheeked baby on my chest. The sequence was disconcerting: the outcome of the process (a baby) so absurdly disproportionate to the blandness of the effort expended.

But with my second baby, the best-laid plans went awry. It was a sticky end-of-summer day in Philadelphia, and I and my shiny belly, swollen with child, were taking a rest in the cool of a darkened bedroom. My husband and three-year-old daughter had just gone out for a walk. I was lying in the bed, trying to block out the noisy struggles of the window air-conditioning unit, when I heard my water break: an absurdly audible squelch-squeak-pop like the bursting of a water balloon. Labor kicked in, and my mind flooded with lurid stories, real and imagined, of how quickly second labors often progressed: babies suddenly popping out in bathrooms, on sidewalks, in taxicabs. My husband had forgotten his phone. Panicked, I hauled up from the bed and drove myself to the hospital.

I careened into the semicircular emergency room driveway, stagger-sloshed out of the car, and made my way dripping in an earth-toned H&M peasant skirt into labor and delivery. But the resident botched my epidural. By the time they realized it hadn't worked, it was too late to try again—"time to start pushing," my doctor said.

The thing about labor pain is it is so big it leaves you speechless. Not: I can't talk, I'm in too much pain. But: This is a mystery; I never knew something this big existed. So big it doesn't even feel like it is happening inside my body; so big it exceeds words, imagination, skin. There is nothing to contain it.

Somehow, I knew instinctively that if I tried to fight—flapped or fluttered or kicked like an animal caught in a snare—I was a goner. I don't know how I knew this, having never taken a birthing or breathing class of any kind, but I did. You couldn't run away from this pain; you had to join it, sink deeper into it, acquiesce to this sundering of your person. The term "labor" is far too active a word, implying volition, effort. This is true up to a point, but at bottom, as Maggie Nelson rightly points out, the labor of childbirth isn't something you do; "labor does you."

Early critics of anesthesia worried that by taking away the biblical curse of pain in childbirth, ether would lead to atheism. "Pain is the mother's safety, its absence her destruction," insisted one outraged obstetrician in the *Edinburgh Medical and Surgical Journal* of 1847. And yet there were those rash enough to administer the offending vapor, forgetting that "it has been *ordered* that 'in sorrow shall she bring forth.'" While the sexism of the remark rankles, the spirit stands: pain is indeed a reminder of finitude, a claim on our obedience to something outside the self. It makes us an offer we can't refuse.

What I learned in labor: to be obedient to the mystery of something that surpasses not only the limits of your physical capacity to suffer, but of your capacity to know. Man is not the measure of all things. There is the measureless, and when you encounter it, you get the vanity knocked clean out of you.

In sickness unto death, in grief, in labor, you brush up against a power that shatters the limits of personhood while still sticking you to the wall and insisting that you bear witness to it *as a person*. This involves a turning inside out of being, a cleaving of the self that, once done, cannot simply and cleanly be undone. Pain smashes all idols, even the one we believed could never be touched: the free, willing, and inviolate self.

It is a deeply human impulse to want to skip quickly to the other side of this sundering of the person that is extreme suffering—our own or that of a dying loved one. "Full fathom five thy father lies; / Of his bones are coral made; / Those are pearls that were his eyes: / Nothing of him that doth fade, / But doth suffer a sea-change / Into something rich and strange," sings Ariel in *The Tempest*. We want to suture back together the person we've lost, to bring closure to their story, and thus to ours. We seek to clothe the mystery that is pain in comfortable human shapes, to exchange suffering flesh for a pearl.

Flannery O'Connor wrote of this narrative instinct toward repair. "There is something in us, as storytellers and as listeners to stories," she wrote, "that demands the redemptive act, that demands that what falls at least be offered the chance to be restored." But suffering, by its very nature, is anti-narrative. Pain resists being plotted as progress. I don't believe that my father is "at peace" or "in a better place." He suffered, and the suffering was terrible, and also terribly ordinary, and now he is gone. When I held his hand and smoothed his wild hair and cupped his face as it settled into death, I saw no pearls where were his eyes.

In the Old Testament the Lord leads Ezekiel into a valley of dry bones. "Son of man, can these bones live?" queries God of the prophet. And he responds: "there was a noise, and behold a shaking, and the bones came together, bone to his bone. And when I beheld, lo, the sinews and the flesh came up upon them, and the skin covered them above." Ezekiel sees a whole army of the resurrected, risen in glory to be led into Israel triumphant.

I don't say that restoration is impossible. But neither should we approach its prospect with what O'Connor once called tartly the "foregone optimism" of our sunny, can-do age. Above all, restoration can never be gained by averting our gaze from the mystery and the pain of incarnation. It is somewhere in this jumble, this shakedown of the self, this sea-change joining sinew to spirit with imperfect joints, patiently stitching word to bone, fragment to fragment, that redemption (maybe) does glimmer.

CORPUS CHRISTI

I took my first Catholic communion in 1973, when I was seven years old. There was no white lace wedding-cake confection with a veil, no patent leather shoes often associated with the rite. Instead, my mother bought me a sky-blue sleeveless Izod shirt-dress that she had picked up somewhere half price and which, in her view at least, was sufficient unto the solemnity of the occasion (or in any case, less vulgar than the whole tiny-bride-of-Christ routine). I looked less like a communicant than an aspiring tennis champ.

Still, I watched as Father Tom reenacted the Last Supper, preparing the body and blood of Christ for our consumption. "Take this, all of you, and eat it. This is my body, which will be given up for you." My mother's parting instructions as I left the pew to go up to the altar: "Don't fidget in the communion line, and don't chew the host." This latter request was easier said than done. Lodged between my tongue and palate, the wafer would all too often stick to the roof of my mouth. Once back in my seat I'd have to surreptitiously scrape it off, sucking the dissolved remnants of God's body off my index finger.

The indelicacy of sinking one's teeth into Christ's flesh had always posed a theological quandary for the Church. Saint Augustine took a politely symbolic view of the eucharist, referring to the bread and wine as visible "signs" only, and thus distinct from the divine substance they signified. Later scholastic theologians, however, doubled down on the literal identity between body and bread and performed ever-more-baroque spiritual gymnastics to account for how, in fact, the sacred substance could be metabolized—chewed, swallowed, digested—without losing its sacred character. So, the question stands: What does it mean to digest God?

In an actual back-and-forth dialogue among tenth century monks regarding the hypothetical case of a church mouse eating leftover eucharist crumbs on the altar, more indulgent theologians gave the creature a pass. Animals, not being rational, could hardly receive the host as a symbol; instead, they gobbled it down as mere foodstuff. Other theologians, however, took no chances; Peter de la Palu insisted that in such a case, the mouse would have to be burned and its ashes returned directly to the earth in order to avoid desecration. Following a similar logic, twelfth-century theologian Alain of Lille recommended that any criminal condemned to death be denied communion in the three days prior to his execution, lest

"Christ, who was still believed to be in the stomach, should be handed over to hanging," killed twice in a species of theological double jeopardy. (That it was less sacrilegious for Christ's body to be excreted rather than hanged seems not to have come up for question).

Thomas Aquinas came to the rescue in the twelfth century with a neat philosophical solution: transubstantiation. Any physical thing was composed of *substance*, Aristotle taught—an essence that subsisted across time—and *accidents*, the myriad incidental shapes and changes that a substance might undergo. For instance, a man might be tall or short, bent or straight, but beneath the accidents of appearance, his essence as "man" persisted unchanged. In the same way, argued Aquinas in the *Summa Theologica*, what is "broken and chewed" in the act of eating the eucharist is not the essence or substance of Christ's body, but the mere accidents of bread and wine. Thus, by an Aristotelian sleight of hand, was God's body saved from the indignity of matter.

I remember my parents trying to explain transubstantiation to me. "You know the Episcopal church, the one with the red door, on Diamond Park?" my father asked me. I did. "They take communion too. But when they do it, they think it's just a symbol. Catholics believe it is *actually* Christ's body and blood." This cleared matters up a little bit, but in the end just led me to conclude that Catholics believed in magic and

Protestants didn't (which, I suppose, is a short explanation of the Reformation).

For a while this knowledge made me feel special, like I had access to a secret superpower denied my Protestant friends. Still, the fact was unavoidable: at the heart of Christian belief lay a rather crude act of ingestion. In his 1890 work of speculative anthropology, *The Golden Bough: A Study in Comparative Religion*, Sir James George Frazer traced the Christian sacrificial meal back to Bronze Age vegetation cults. The "savage mind" of early humanity, Frazer hypothesizes, pictured the change of seasons as a drama of life and death, keenly aware that "the same processes which freeze the stream and strip the earth of vegetation menace [man] with extinction." Through ritual reenactment of the death and resurrection of a god of nature, early agricultural societies hoped to secure the annual return of life-giving crops.

In Frazer's genealogy, the earliest avatar of this sacrificed nature divinity is the Assyrian god of grain, Adonis, young lover of fertility goddess Ishtar. When Adonis is killed by demons and dragged to the underworld, Ishtar descends to his rescue, and the earth is left barren by her departure. Only when the couple reascends to earth does life return to vine and womb. In the yearly Babylonian festival memorializing Adonis's death and resurrection, women mourners worshipped an effigy of him, patched together of honey, wheat, and flowers, while bewailing the world's barrenness in his absence: "Her lament is for the woods, where tamarisks grow not," runs one funereal hymn. "Her lament is for the depth of a garden of trees, where honey and wine grow not. . . . Her lament is for a palace, where length of life grows not."

Other versions of the myth were bloodier yet. In one telling, the Phrygian goddess Cybele, lovesick for the fair young shepherd Attis, struck him with a fit of madness to prevent his marriage. When the frenzied youth unmanned himself and bled to death beneath a pine tree, the grief-stricken goddess made a deal with Zeus to resurrect her lover. The cult of Cybele, widely popular under the Roman Empire, celebrated an extended festival during the vernal equinox. A pine tree, wreathed in the violets said to have sprung from Attis's blood, was carried into Cybele's sanctuary. Whipped into an ecstatic frenzy, the cult's clergy slashed their bodies with potsherds and knives, spattering the altar and the sacred tree with their blood, while the more zealous castrated themselves in imitation of their god. The god-in-effigy was then buried and, on the morning of the third day, when the tomb was opened, Attis was found to have risen from the dead. Initiates into the mystery cult of Attis, Frazer suggests, partook of a sacramental meal by "eating out of a drum and drinking out of a cymbal," prefiguring the Christian supper.

Christianity was, then, no new thing under the sun when it came onto the scene in second and third century Rome. Frazer discusses the importation of what he calls such "Asiatic" mother goddess cults hailing from parts East—Christianity included—with barely concealed disgust. "The ecstatic frenzies, . . . the mangling of the body, the theory of a new birth and the remission of sins through the shedding of blood, have all their origin in savagery," he claims. Such cults sapped the manly Roman vigor of the Western empire, and it was only with the revival of Roman law and Aristotelian philosophy at the close of the Middle Ages that "the tide of Oriental invasion had turned at last."

At pains to explain how the more sophisticated Roman citizens could have fallen for such a crass Oriental bait and switch, Frazer concludes that the "true character" of these rites was often craftily "disguised under a decent veil of allegorical or philosophical interpretation." And, indeed, Anglo-American religious history has never tired of pointing out the childish, savage, and overall distasteful nature of Catholic literalism as embodied in the eucharist. As the Reformation historian Preserved Smith noted in a 1918 essay on "Christian Theophagy," Catholics fail to heed Cicero's civilized skepticism regarding the gods and their associated sacrifices: "'When we call corn Ceres and wine Bacchus,' Cicero mused in his essay on religion, 'we use a common figure of speech: but do you imagine there is anybody so insane as to believe that the thing he feeds on is god?'" As Cicero to the "savage" and overly literal worshippers of Bacchus and Attis, so prim nineteenth-century Anglicans to those wearisome Catholics too dumb to recognize a metaphor when they saw one.

Growing up in Syracuse in the 1940s and '50s, my mother butted up against these prejudices again and again: she and her Irish parents were credulous peasants, trapped in a backwater eddy of superstition, throwing sand in the gears of civilization. "Mackerel snapper," snickered Protestant school mates. (And in fact, my grandmother did cook fish for supper on Fridays.)

Yet the Catholic insistence that one eats God's body ("broken and chewed," to cite Aquinas) not as symbol but as literal enzymatic and chemical fact—as metabolic feat—helps make sense of the gospels' otherwise baffling obsession with broken and suffering bodies. There is no more visceral way to experience the fact of our shared flesh—to understand that our participation in the other is literal, not figurative—than to come face-to-face with the physical vulnerability of another's body. Again and again, the historical Jesus of Nazareth scandalized his community by provoking precisely this gut-level reaction of outrage and disgust at our broken one-ness. He walked around Galilee and Judea touching any and every person who might be considered low or unclean: rotten flesh, empty bellies, blasted eyes, dirty feet, bleeding prostitutes. Instead of cursing the broken bodies, he consecrated them. More confounding still, he insisted there was no difference between these bodies and his own. "Verily I say unto you, Inasmuch as ye have done it unto one of the least of these my brethren, ye have done it unto me." And to this outrageous fact of participation in, and responsibility for, a larger order of being, he gave an unprecedented name: he called it *love.*

As part of the process of getting confirmed in the Church in eighth grade, I went on a service trip to a poor rural pocket of West Virginia. We split up into groups and I went to visit a homebound elderly man, bringing him food and, perhaps more importantly, an hour of company. My memories of the place are blurry: I see a single dark room, sparsely furnished, with a bare floor reflecting tepid sunlight from the front window. Of the old man himself, sitting on a sofa, I retain only this: that the flesh of his face was pale, pasty, crumpled looking. There was something larval about his presence, indistinct and folded in on itself, like the paint smudge of a face in a Francis Bacon painting. At the sight of his collapsing flesh, I felt pity and disgust.

During the course of this visit, the man took down from a shelf a gray object the size of his palm to show us: a tiny box, complete with a lid, carved out of the tooth of a sperm whale. It was not a fine or polished bone sculpture; it appeared rough and rusty and dirty as if still trailing bits of the whale maw out of which it had been plucked. Still, we politely pretended to wonder at it.

The visit has stayed with me, perhaps because it was the first time I understood how hard it is, counter-intuitive even, to "love one's neighbor" as my religious training was asking me to love this man. It went against the will, against all desire and inclination: on the contrary, it induced a sharp spasm of creaturely revolt. The service worker who brought us on the visit was serene, patient, efficient. For her, this rite of bearing witness to our shared imperfect flesh had become reflexive: like muscle memory, like prayer. *This* was the meaning behind the rite of swallowing God's body. *This*—in this dim room, admiring an indecent hunk of whale's tooth in the old man's shaky palm—*this* was what love looked like, and it terrified me.

I was a very poor science student in high school, barely scraping by in physics and chemistry, but I loved biology—perhaps because it is such a literally incarnational science. I dissected a worm: slid a scalpel along its rubbery length, then pulled back the flaps and pinned them to the bed of black wax lining a dissection tray. Inside, its organs matched point for point the diagram in my textbook: gizzard, gonads, intestine—a tiny hidden universe laid to light. Next up, a frog. I remember the feathery green-brown lobes of liver; the duodenum a shiny, smooth kink of flesh leading away from the pillow of stomach. The organ looked like it would yield to my scalpel but instead resisted, as if one were cutting into a rubber ball, or biting down on a licorice braid. Rubber, bounce, recoil: I can still feel in my right hand—ghost of a muscle-memory—the surprising resistance these inert bodies showed my instrument, as though instinctively protecting their inwardness.

I remember, too, a poster I made illustrating the process of phagocytosis: an amoeba eating a paramecium, shooting out its jelly-like appendages—a primitive, make-shift mouth—to encircle its prey before digesting it, stripping it for parts then ejecting the waste. This is what heterotrophs—"other-eaters," from amoeba to humans—do: whether their mouths are formed by cytoplasmic claws or fully evolved teeth and tongue, they eat other living or once-living things and incorporate them into their own flesh and blood, metabolizing foreign substance into the substance of the self.

In the nineteenth century, researchers in the nascent science of physiological chemistry set out to give a complete account of the transformations undergone by matter as it passed through a living organism. Termed *Stoffwechsel*—stuff-changing or matter-changing—in German and *metabolism* in English, early metabolic theories compared food to fuel, and the body to a stable fuel-burning machine on the model of the combustion engine: matter goes in; energy and waste come out. In 1839, Jean Baptiste Boussingault fed a cow a carefully measured portion of potatoes and hay, collected the animal's resulting milk, feces, and urine, and by comparing "the food consumed and the products rendered," tabulated a chemical analysis to measure gains and losses in nitrogen, oxygen, hydrogen, carbon, and salts.

Still, Boussingault's precision instruments notwithstanding, the cow's inner workings remained a bit of a black box. These studies often posed a philosophical as much as a chemical conundrum, making early physiologists the unlikely heirs of those medieval theologians who had passed their days puzzling out the mystery of the eucharist: How to account for the transformation of one kind of matter into another? How could one tell the difference between substance and accident in this most mysterious of all transubstantiations, elements roiling in the dark cavity of animal gut and lung?

Eventually, the introduction of isotopes into these experiments, through which scientists could follow the metabolic trajectory of individual molecules, revealed eater and eaten to be far more intimately bound than previously imagined. Food wasn't just fuel passing through the engine of an organism, it turned out; its molecules were swapped out with the engine works themselves. Foreign *Stoff* was metabolized into the very architecture of the self, and what was once self, broken down into foreign *Stoff* to be sloughed off in turn, substance and accident hopelessly entangled.

Perhaps, then, the Bronze Age worshippers of Adonis and Ishtar, with their patched-together idols of honey and wheat and flowers, and the nineteenth-century physiologists, with their combustion tanks and scales and inky balance books, were on the track of the same truth, albeit expressed in different idioms. For what, after all, is metabolism, this endless loop of eating and being eaten, if not the chemical expression of the most ancient myth of all: the cyclical death of the earth (god); its dispersal, ingestion, and rebirth; the original splitting that consecrates our oneness; the cosmic molecular arc that binds all life?

Once, my catechism teacher told the class that even when taking communion on an empty stomach, she "felt full" afterwards. Then she smiled beatifically. When I reported this possibly miraculous circumstance to my mother, she rolled her eyes and dismissed it as "unlikely." And besides, she continued, "What does God care whether she feels full or not?"

Our participation in a larger body—of a church, yes, but also a community, a species, a planet—is a truth that is true regardless of what we "feel." "When I ask myself how I know I believe," Flannery O'Connor once wrote, "I have no satisfactory answer at all, no assurance at all, no feeling at all." It has been almost thirty years since I've taken communion or even stepped inside a church. Yet one lesson my overly literal Catholic upbringing taught me remains: that our shared flesh and fate is not a feeling or a choice or even a belief but a fact, *the* founding fact, measured against which all other facts are as nothing.

There was a side table in the living room of the house I grew up in where my parents kept a cluster of Catholic artifacts picked up in their travels through Mexico, small statues and portraits that had once adorned devotional shrines. A faded painting of Saint Jerome on tin, showing him seated in front of his cave in a red cloak, head pensively cradled in an open palm while a curly-haired lion lies at his feet. A blackened wooden crucifix with hanging Christ, his tiny face pinched in pain, with real nails piercing his hands and feet and a miniature crown of thorns. A painted plaster saint-in-robes whose hands had long ago broken off, such that the empty sleeves of his raised arms appeared to hold forth the gift of nothing. All these miniature icons in various stages of decay did little to counter my intuitive sense that to be a believer in this Church was inescapably tied up with the fragility of our bodies.

It wasn't until much later in life—after I had had more experience with my literal body being cracked up and cracked open—that I'd realize it wasn't brokenness per se that I'd been taught to hold sacred. Rather, I knew these ragged-edged household gods were meant to serve as a reminder that we are all shards of a larger whole that transcends our knowing. That the edge of our skin is not an end but a beginning; not a closing off but an opening up; the breach that binds us each to each.

OUTIS

I spent the better part of the 1990s as a Comparative Literature graduate student at Princeton University. To anyone who was a Comp Lit grad student (or even Comp Lit adjacent) during the '90s, the following list of things I did during this time will appear entirely predictable: I wore an oversized black leather biker jacket with miniskirts, rain or shine; I read and pretended to understand Hegel's *Phenomenology of Spirit*; I drank too much coffee, too much vodka, and I fell in love with a post-modernist architect (who also dressed mostly in black).

Something else I did, not out of character for a '90s Comp Lit grad student but not quite as common as black leather jackets and Hegel: I had an old-fashioned nervous breakdown. The kind where my parents had to drive seven hours from Western Pennsylvania, trundle my catatonic self into the back seat of their Toyota Corolla, then turn around and drive seven hours back.

When we arrived home, my mother propped me up on the lumpy red Victorian sofa in our living room, swaddled me in blankets, and fed me tea. Then she sat down and waited for me to come back from wherever it was I had disappeared to.

I had been afraid of going crazy since the age of thirteen. It started with intrusive thoughts, followed by counteracting tics. One day, leafing through some books in my grandmother's house, I fell upon a history of Joan of Arc. I read about her being burned at the stake, and suddenly a still, small voice in my head said, "What if *you* were burned at the stake?" (An alternative version of this thought was, "What if you were hanged as a witch?") And that was it.

Suddenly, the possibilities of burning and hanging lurked everywhere: in my beat-up green eighth grade world history textbook, with its section on the Salem witch trials; in red, orange or yellow clothing, whose flame-adjacent energy might spark a fiery demise; in a suspiciously noose-like length of garden twine; in the button-down shirts my mother bought me at a cut-rate outlet store with neck sizes printed on the tags. Coming into contact with or even thinking about the forbidden objects touched off a frenzy of compulsive counter-charms aimed at warding off this grim fate: handwashing, wall-tapping, silent word-repeating—sometimes a magic combination of all three.

Obsessive-compulsive disorder is maddening for its almost prankish variability. It is a shape-shifter, working by a logic of association and contamination that knows no stopping place. One nineteenth-century psychologist observed of his patients' tics and compulsions that they constituted a kind of "associative mania": never stable, they spread like "an oil stain." In the meantime, the sufferer strives to salvage tiny rafts of safety on which she can float, safe objects and a fleet of taboo-canceling rituals. But the malady's perversity is such that just when you think you have a fix on every danger and how to avoid it, a new risk crests the horizon.

You can sense it coming, like an aura before a migraine. I remember one day in junior high searching for something to wear to school, choosing from an ever-dwindling pool of clothing as first one, then another, became fire and rope contaminated. I pulled out a T-shirt that, up until then, had passed as safe: red, orange, green, and white stripes. The white and green had, hitherto, worked as a counterforce to the red and orange. But as I started to pull it over my head, I realized, with a feeling like a lump of ice being drawn down my spine: green, if you broke it down, was composed of yellow + blue, and orange was composed of red + yellow, giving the forces of fire a clear numerical superiority over the forces of anti-fire (4:2 or 3:2, depending on how you calculated it). In either case, these were statistical odds I would never be able to risk. I remember sitting on the cold basement steps, clad in just my underwear, sobbing into a pile of folded laundry.

Then, in eleventh grade, I took a psychology elective and became convinced I would develop schizophrenia. The primary indicator of schizophrenia, according to the textbook, was the occurrence of auditory hallucinations. Hearing voices was a hands-down, telltale sign. I became hyperaware of my mental landscape, tirelessly scanning for the faintest sign of a foreign voice batting about my brain, an outsider come to usurp my place and seal my exile in the kingdom of the crazy.

Falling asleep was the worst, when the natural quiet of the night swelled to deafening, pregnant with the possibility that at any second a voice might explode inside my skull. I strained my ears, surveilling the border between inside and out, panicked that the distant hum of an engine or the faint scratching in the walls—squirrels in the eaves—might in fact be coming from inside my head. "Did you hear something?" I'd hiss to my sister across the dark void separating our twin beds, fingers white-knuckling the comforter. When she'd mumble back in the affirmative, the flood of relief that washed through my body—no, not crazy, not yet—was enough to make me weep. But the next morning, as soon as I opened my eyes, it would start all over again.

And while for reasons beyond my ken I was granted a reprieve from obsessive thoughts for the four years I was in college, they returned during my first year of grad school. It started one day as I was taking a shower in the bleak communal bathroom of the graduate dormitory. As I soaped up my arm in the splintery spray, wet skin and water lit up like sparks in the glare of the merciless overhead lighting, I suddenly froze. It started as less a thought than a prickle along my spine, an impression trying to gather shape somewhere in the back of my brain. But before I could stop it, at first slowly and then all in a rush, I looked at my hands and didn't recognize them as belonging to me. A thought formed: "Are these mine?" *Maybe—but who is "me"? Who is the person who just asked, "Are these mine?"*

I (*Who is 'I'?*) scrambled to get ahead of the infinite regress of *I*'s suddenly spawned in my head, but knew in advance in the brain, in the gut, in the electric panic surging up and down my body, that there was no end. There was a lag in my being, like I was either a step ahead or a step behind myself, but nowhere coincident. I wasn't doubled, exactly—not twinned or *two*—but neither was I *one*. Caught between self and not-self, I was no one.

From its earliest study, natural philosophers and medical writers treated what we today call obsessive-compulsive phenomena as a subset of melancholy, a state of deep, brooding sadness attributed to a humoral imbalance in the blood. Robert Burton, whose 1621 *Anatomy of Melancholy* is a sprawling, encyclopedic treatise on the disorder, follows the second century Roman physician Galen when he traces attacks of melancholy to faulty "concoction" of food in the stomach and liver. The "windy vapors" emitted from improperly digested meats rise to the brain, tainting it with a sooty, blackened aura: "As a black and thick cloud covers the sun, and intercepts his beams and light," Burton explains, "so doth this melancholy vapour obnubilate the mind, enforce it to many absurd thoughts and imaginations." This corrupted humor, wandering "to and fro in all creeks of the brain," gives rise to "superfluous and continual cogitations," "divers monstrous fictions." Burton was, perhaps, the first diagnostician to note the excruciating self-consciousness of the obsessive, methodically aware of his madness: "In all other things they are staid, discreet, and do nothing unbeseeming their dignity, person, or place." Burton notes, "this foolish, ridiculous, and childish fear excepted." Yet against it they are powerless, and the obsession "continually tortures and crucifies their souls, like a barking dog that always bawls, but seldom bites."

Possible remedies ranged from herbal drinks and potions to stone amulets. Red garnet, "a precious stone so called, because it is like the kernels of a pomegranate," Burton counsels, "if hung about the neck, or taken in drink, much resisteth sorrow." In more stubborn cases, Burton advises, "'Tis not amiss to bore the skull with an instrument, to let out the fuliginous vapours." Though here some experts disagree, claiming melancholy "too stiff a humour and too thick . . . to be evaporated": Drill at your own risk.

Nearly three hundred years later, nineteenth-century medical men in the emerging field of psychology still classified obsessive thoughts and tics under the broad umbrella of melancholia, both conditions characterized by self-absorption and doomlike funk. The humoral theory of disease had been largely discarded, replaced with a vision of the nervous system as electrical network on the model of the newly invented telegraph. Psychological maladies were chalked up to a glitch in the system, a block in the flow of nervous currents, like a garbled or undeliverable telegram. In his 1895 treatise *The Pathology of Mind*, English physician Henry Maudsley compared the patient in the throes of an obsession to a machine gone haywire, "locked in impotence by the cramp-like action of the morbid thought-tract and the paralysis of the rest of its machinery." Indeed, the peculiar torture of the obsessive, according to Maudsley, is that the sufferer, "all the while perfectly conscious of the [ludicrous] character of the idea" tormenting him, is yet unable to act or think otherwise. Like Burton before him, Maudsley can only note with perplexity the way obsessive thoughts "come and stay there against the [patient's] will, a haunting horror, a maddening torture," reducing the patient to a "reflex organic machine."

⁓

And so, after my parents brought me home, I took up residence on the Victorian sofa, a fitting enough perch for a twentysomething girl paralyzed by nervous collapse—or what Victorian physicians would no doubt have called neurasthenia, "weak nerves"—to find herself.

The specific form my obsessions chose to take in my twenty-fourth year was what early twentieth-century French psychologist Pierre Janet catalogued under the rubric of "Feeling of Incompletion in the Perception of One's Person," or what today would be called depersonalization disorder. Janet quotes from his patients experiencing this peculiar psychic state: "I can't seem to arrive at the unity of my person," one of his patients complained; "I can't seem to reach myself." "I was listening to my voice as if it belonged to someone else," explained another patient. "All the while I knew it was my voice, but the 'me' I heard speaking felt like a lost self, an old self, far away." All described a feeling of spatial or temporal lag, which led Janet to propose a parallel with the much more common mental phenomenon of déjà vu. Except instead of a single memory being doubled, it is the bearer of memory herself who feels an uncanny sensation of doubling or repetition, at once more-than-one and less-than-one.

What I remember most from that period was the terrifying impossibility of communicating my plight to anyone. In trying to convey to concerned onlookers (my parents) that my person had absconded, I could do so only by borrowing the vocabulary of personhood. Forced to lay claim to an "I" and "me" that were precisely what no longer existed, any attempt to give words to my experience immediately negated it. "I" was just a floating thought, resistant to grammar, to all yoking of a subject to a verb or an object, a perpetual idiot stutter. Every time I uttered the syllable *I*, it tripped off a cataclysmic cascade of panic inside my brain. I would have to hold very, very still, striving to think of nothing, until the feeling receded.

My mother arranged for me to see a therapist, about whom I knew only that she was a former Miss Crawford County beauty pageant winner (a distinction she shared with Sharon Stone, who took the crown in 1976). My therapist wore dark pantsuits with high heels and had, naturally, a dazzling smile. She was very smart and kind, and our weekly meetings made me feel, if not better, then at least a little less despairing. She convinced the only psychiatrist in town—a brusque, dismissive man who, once I had gathered the courage to confess my symptoms to him, immediately told me I was "probably" schizophrenic—to prescribe a new drug that doctors were finding very helpful in treating anxiety and depression. It was 1991; it was called Prozac. I began swallowing my daily dose.

In his 1912 book *The Life of a Caterpillar,* French entomologist Jean-Henri Fabre presents the curious case of *Thaumetopoea pityocampa,* or the pine processionary caterpillar. During a key phase of its life cycle, as many as three hundred of these tiny creatures line up head to rump, single file, and follow a female leader to a pupation site, where they then bury themselves and form cocoons. Curious to test the strength of this marching instinct, Dr. Fabre placed the female leader on the rim of a glass to see how long it would take the pack to realize they were moving in a circle. They walked around and around the rim of the glass for more than a week. They walked until they died. "The headless file has no liberty left, no will," Fabre observed of the weary marchers. "It has become mere clock-work."

It is the closest I can come to describing the way I experienced time during my nervous breakdown. There were no markers to measure out space and time, no beginning and no end, no way to *will* a break in the endlessness and start anew. "The obsessive patient," Pierre Janet once observed, "never arrives at a conclusion, can never 'settle accounts,' and so exhausts himself in a labor as interminable as it is useless." The time of obsession is mythic time, the maddening, circular time experienced by the shades in Hades: the Danaids condemned to carry water in a punctured pail; Sisyphus to roll a stone uphill; Odysseus to kiss his mother, only to find her vanishing at his touch.

And then one day, two weeks into taking Prozac, the will-less, headless caterpillar circle march that was my interior world suddenly shifted. As if the queen caterpillar had, for no particular reason, taken a leap off the edge of the glass and started inching in the direction of the nesting ground, pulling the rest of the train in her wake. To try another analogy: I emerged out of my illness the way a crystal takes shape out of a saturated solution. Stir sugar in water and watch it dissolve until the mixture is evenly combined, a kind of liquid cloud. Tip it one tick in any direction and the sugar molecules will start to precipitate out: slowly at first, a few molecules clustered on the wall of the glass until finally the sharp, clean edge of a right angle juts out of the monotone murk. Out of the suspension—featureless expanse, void of trace or line or mark or thickening by which one might get one's bearings—a crystal quickened.

It happened so gradually that, at first, I didn't realize it was happening at all. Once I sensed it—mind-stuff pulling taught, a crease in chaos—I was afraid to believe in it. I side-eyed it gingerly, this spot of color against the gray, afraid that if I peered too closely the tiny cluster of a self that had interposed between the void and me might disappear again.

But instead of slipping back into the widening gyre of my panicked brain, I took one step forward, then another. I got off the sofa. One day I started baking bread—punching down the yeasty dough, waiting for it to rise beneath a damp cloth set on the floor in a patch of sunlight, spritzing water into the hot oven to create (so Jacques Pépin assured me) a crispy crust. I deboned a *gigot d'agneau* I somehow procured at Kroger and baked it with salt and rosemary. I made Julia Child's salmon soufflé, breaking through its crisp pink-bronze dome with the gleaming edge of a fish knife, proffering it to my astonished parents. The physicality of cooking—chopping, blending, stirring, beating, blowing, tasting—offered my battered mind a place to land, a solid perch I could trust to keep and hold its weight. Bit by bit, I felt myself pulled back into the forward-pushing time of the living.

In a downtown composed of little more than two stoplights, a defunct movie theater, and a Dunkin' Donuts, I noticed one day that someone had optimistically launched a catering business in one of the many empty storefronts. Sarah of Sarah's Café was short and perfectly round, with a soft doughy face, throaty giggle, and gunmetal-gray hair sculpted into a perfect Aqua Net bouffant. She wore green eyeshadow. Her husband was a preacher with piercing blue eyes and a bum leg that forced him to hitch up his left hip with every step he took. Despite the bad leg, he was immensely strong and did all Sarah's heavy lifting and hauling. I asked if I could give them a hand with the catering, and they welcomed me aboard.

It was springtime. We baked cakes together. I would load my car with Sarah's sugar-crusted confections—even the savory dishes were somehow sweet—and drive out into the countryside, to a birthday party or a first communion, the scent of freshly mowed grass and wet earth and daffodils warmed by the sun blowing soft on my cheek.

We no longer believe in the humoral theory of Galen and Burton or the mechanistic psychology of Maudsley and Janet; we no longer attribute obsessive-compulsive behaviors to the ill-cooked contents of our livers sending sooty fumes up to sicken the brain nor to poorly distributed nervous energy that might be coaxed back into proper circulation by the skillful application of magnets to flesh. Yet the biological underpinnings of obsessive-compulsive thoughts and behaviors remain, to this day, largely a mystery.

A 2018 University of Michigan study using MRI brain scans recently discovered one possible lead: failed messaging between the cingulo-opercular network, an area of the brain responsible for monitoring errors, and the prefrontal cortex, the area responsible for acting on that information and correcting course. In responding to errors during a brain scan, OCD brains showed extra-firing capacity in those sections of the brain where errors get flagged, but reduced capacity in those areas required to correct them. According to authors of the study, the OCD brain gets stuck in a "loop of wrongness," leaving the sufferer powerless to exit it.

A bioelectric quirk of the brain, maybe—but one that revealed to me, beyond the sphere of synapses and molecular misfires, something about the very nature of this self that we are so quick, in the Western tradition, to equate with action, freedom, control—not to mention progress and forward movement. For a long time, I couldn't bear to look back on that period of my life, averting my gaze whenever my mind's eye chanced to wander there. But then, the experience of depersonalization is such that it can never become the steady object of a knowing, measuring gaze. It has its very essence in evasion, splitting, looping: the measureless.

It is perhaps not by accident that one of the most penetrating descriptions of depersonalization was written by Emily Dickinson, herself diagnosed with neurasthenia in 1884. One of the many poems she dedicated to what she called her soul's "bandaged moments" proceeds by listing, with surgical precision, everything this particular experience of psychic pain *was not*. "It was not Death, for I stood up / And all the Dead, lie down," she opens. "It was not Night, for all the Bells / Put out their Tongues, for Noon." The poet presents a litany of things this amorphous "it" was *not* like—frost, fire, midnight, space—before lighting upon a closing analogy: "But most, like Chaos—Stopless—cool / Without a Chance, or spar—/ Or even a Report of Land—/ To justify—Despair."

This experience of being ousted from one's person can thus only be approached through the *via negativa* of subtraction, and in this, I recognized only many years later, an encounter with the void is indistinguishable from an encounter with God as described in mystical traditions (not that I have had one—I have not—but I know what it is supposed to look like). "There shall no man see me, and live," God says to Moses in the Old

Testament, and for that reason he hides Moses in "a clift of the rock" as he passes by: "And thou shalt see my back parts: but my face shall not be seen." So, too, in the Upanishads, the pilgrim seeking the nature of the All-Being or Brahman can do so only obliquely, through an endless process of subtraction: *neti, neti*—not this, not that.

The *I* is a grammatical impossibility in the syntax of the encounter with the Absolute, whether void or plentitude, nothing or everything. And in some sense there is no real difference, as Ralph Waldo Emerson, in his most pixelated states of communion with nature, knew: "Standing on the bare ground, my head bathed by the blithe air, and uplifted into infinite spaces, . . . I have become a transparent eyeball; I am nothing; I see all; the currents of the Universal Being circulate through me; I am part and particle of God." Both encounters—void, all—involve a form of rapture, the absconding of the *I*, and both encounters disable the familiar relations of knowing and choice that characterize interactions with objects in our day-to-day dealings with the world.

After a semester and summer spent at home, I went back to grad school. I took up *The Phenomenology of Spirit* where I had left off. I added Dostoevsky's *The Double* to my reading list. In short: I reprised the onward-pushing march that is my life.

But I have never forgotten the stopless time when, as Dickinson writes, I lay awash in chaos, without a chance or spar. When even despair, as a fixed and chosen relation to the outside world around which one might build a self, was out of reach. It left me with an abiding awareness of the temporary, make-shift nature of this *I*—splinter of a spar, clift of a rock—that at any given moment holds me safe.

BODY MAP

The act of giving birth has always been a mystery, starting with the basic mystery—as old as the hills—of how the world ever got from one to two in the first place, from unity to multiplicity. Birth is akin to a riddle or a magic trick: one person goes into a room; two people come out. And perhaps magic tricks so often involve numeric befuddlement for precisely this reason. A charming assistant sawed in half, rabbits pulled from an empty hat, marbles vanished from beneath swapped shells: so many echoes of that most archaic number puzzle of all, the original head-scratcher.

But the peculiar multiplication that happens in birth goes beyond the bare fact of replicating cells and bodies. Rather, this event permanently alters the birthing body's relationship to number and extension in space more generally. First there was one, then there were two—but also still one. Wait, am I one or am I two? Am I missing anyone? Once you have a baby you find yourself continually checking and re-checking your roster, counting heads. Somehow the way things added up before no longer holds. There are always going to be too many or too few units, a perpetual remainder or deficit.

Pythagoras, who believed the whole cosmos could be mapped through numbers and their harmonious arrangements, knew that one and two—what he called the "monad" and the "dyad"—are not proper numbers at all. Numbers only become predictable—that is, capable of pattern—once you get to three; before that, one and two dance around each other in circles, incorrigible tricksters, refusing to make a clean start. "The principle of all things is the monad or unit," records Diogenes Laertius of Pythagoras's cosmogony, and "arising from this monad the undefined dyad or two serves as a material substratum to the monad." From this original meeting of form and matter "spring numbers; from numbers, points; from points, lines; from lines, plane figures; from plane figures, solid figures," and so on and so forth until the whole cosmos is set spinning on its three-dimensional axis replete with flowers and birds and temples and stars. And, of course, babies.

In the hospital after my first child was born, after she had been cleaned and wrapped and I had been cleaned and wrapped and my husband had gone home to sleep, the nurses tucked me into bed and placed my swaddled daughter in the crib beside me like a tiny Christmas package. As I fell in and out of sleep, I kept jerking awake with the panicked feeling that I was forgetting something. I reached out, grasping at air, not knowing what precisely I was searching for but convinced that I had left something terribly important behind and was frantic to have it back in hand. The newborn in the bassinet was both it and not it, related to but not the whole of what I sensed was missing. The shape of this hallucination—not being able to account for all my parts—shifted over time into a recurring dream (one that still haunts me). In the dream I have a baby in my care—sometimes it is one of my own waking-life daughters, sometimes a stranger—when I notice that the child has started to shrink, becoming smaller by almost imperceptible degrees, regressing to the size of a kitten, a quarter, a pea, a pin, before disappearing altogether with the soft pop of a soap bubble.

During the dream, there comes a moment where I wonder aloud at what point, in the overall disappearing process, my baby will cease to be a person, and thus cease to be my concern. By what ruler, what compass, will I know when the line separating being from nonbeing has been crossed? The madness of the dream is that the child's hold on me multiplies in inverse proportion to its vanishing; the dream ends when I realize that there exists no smallness so small, no point so infinitesimal, no fraction so fractional, that this being won't continue to act as the center of gravity around which my own clump of matter orbits, in perpetuity. This impossible body's negative extension in space fixes me fast.

And so: after giving birth, wherever you are, there you are not, quite, or not quite *all*. There is a noncontiguous fraction of you extended in space, another self outside yourself, to which, from now to eternity, you are joined. Even if one or the other or both of you should disappear: absence or distance, even the ultimate distance of death, alters not one jot the ghostly geography of this new personhood.

This was, for me at least, not at all (or not at first) a warm fuzzy feeling of oceanic oneness with my offspring. On the contrary: the dawning realization that I had a secret sharer of my being filled me with a cold thrum of panic. Being coincident in your person at all points in space means that, should you so choose, you *could* just disappear. You *could* fall off the face of the earth, wipe yourself off the map. "I have been half in love with easeful death," writes John Keats of melancholy's opiate temptation, and indeed, as a lifelong depressive, I know whereof he speaks.

Once you have a child, a piece of your being exists outside the borders of your body proper, and Keats's impulse to "cease upon the midnight" ceases to be an option. You are, quite literally, grounded—stuck *here*.

Silas Weir Mitchell, who assisted in battlefield amputation surgeries during the Civil War (and, afterward, worked at the so-called "Stump Hospital" for amputee veterans in Philadelphia), was the first to coin the term "phantom limb" to describe his patients' experiences. In his 1866 medical treatise *Injuries of Nerves and*

Their Consequences, Mitchell devotes the final chapter to "Neural Maladies of Stumps." "Nearly every man who loses a limb carries about with him a constant or inconstant phantom of the missing member," Mitchell writes, "a sensory ghost of that much of himself, and sometimes a most inconvenient presence." Indeed, many of his patients reported a more certain and acute consciousness of the missing limb than of the remaining one. "If . . . I should say I am more sure of the leg which aint than the one which are," one patient confided ruefully, "I guess I should be about correct."

The amputees tended to feel most intensely the ends of the amputated limbs—hands and feet—with quite clear perceptions of ghost thumbs, fingers, and toes wiggling in space. One baffling constant in Mitchell's medical testimonies were reports that the missing hand and foot, in lieu of occupying their pre-amputation placement in space, were felt instead to be located much nearer to the trunk. For some, the missing member by degrees and over time crept ever closer to the body "until it [touched] the stump, or [lay] seemingly in its interior,—the shadow within the substance."

Mitchell understood intuitively what twentieth-century neuroscience would confirm: through a laborious process of trial and error, our brains construct a neural shadow-body upon which our substance-body depends in order to move about in space. "We are competent in health, even with closed eyes, to know where and how far removed the hand may be at any moment,"

Mitchell reasons, "and this knowledge is the result of long-continued and complicated sensory impressions, ocular, muscular, and tactile." Today, neuroscientists call this unconscious body knowledge *proprioception.* Proprioception is what accounts for the fact that we can reach out to pick up an object in space without having to visually verify our hand's movement; it is also what accounts for the persistence of the *feeling* of extension in space even in the absence of actual extension, as in the case of phantom limbs.

It wasn't until neurosurgeon Wilder Penfield conducted a series of experiments on locally anesthetized brain surgery patients in the 1940s and '50s that a clearer sense of how the brain constructs a body map, or "body schema," came into focus. By touching specific regions of the patients' brains and recording where the touch was felt on the surface of their bodies, Penfield discovered two narrow strips of tissue running vertically from top to bottom on each side of the brain that contained, in miniature, a "map" of the patient's entire body. At the top of the brain, near the crevasse separating the two hemispheres, is the cortical area responsible for producing sensation in the genitals; next to it, the feet, legs, trunk, hand, face, lips, and thorax in order: a tiny upside-down mannequin with its pieces jointed together (slightly) out of order. "The map is not the territory," semantics scholar Alfred Korzybski famously declared, and indeed it is not: phantom limbs testify to the persistence of the shadow body even in the absence of its corresponding substance—a ghost map to a vanished terrain.

Mitchell wrote his medical treatise at the height of the Spiritualism movement in the United States, when the recent technological wizardry of the telegraph and telephone breathed new life into the age-old dream of communicating with the dead. These machines as conduits for ethereal messages sent from distant bodies cast sudden doubt on the exhaustiveness of Euclidean geometric space: maybe you didn't have to see to believe. It is this metaphysical questioning that comes into play in a short story Mitchell published anonymously in the *Atlantic Monthly* in 1866, "The Case of George Dedlow."

The narrator is a Civil War veteran and triple amputee who has set himself the task of recounting the "metaphysical discoveries" to which his unfortunate experience has led him. An avid amateur scientist, he notes with interest the way the loss of "four fifths" of his weight and "one half" of his skin surface has altered his metabolism: a mere fraction of his former caloric intake now sates his hunger, and his heart pumps forty-five beats to the minute in place of its former seventy-eight. But of more interest still are the psychological changes wrought by his loss of body mass. "I found to my horror that at times I was less conscious of myself, of my own existence, than used to be the case," he confesses. "This set me to thinking how much a man might lose and yet live."

But fractioned bodies and phantom limbs are not unique to amputees. "Indeed," writes neuroscientist V. S. Ramachandran, "even in individuals with intact limbs, the body image is highly malleable; one can lengthen one's nose, or even project one's sensations onto external objects." One laboratory experiment involves having a patient watch a table surface being stroked while their own leg is stroked beneath the table. If the strokes happen in unison, the patient will begin to feel the touch sensations on the leg as if they were coming from the table surface. If the experimenter then hits the table with a hammer, the patient will register a strong galvanic skin response, "as though the object was now part of [their] body."

We like to imagine the integrity of the body envelope, closed up on itself and tightly sealed: I am a single, a unit, a "one," a monad. Yet the fact that the body map can change shape—expand, retract, stretch like silly putty—should make us question our faith in the spatially compact and the uniform. If my brain can induce me to feel an absent limb as present or the surface of a table as part of my body or my child's body as an extension of my own, what other beings might I feel as contiguous to me, shadow of the other tucked into my substance?

When the kids were young, we lived in Philadelphia. In the muggy summers, when the air hung hot over the city like a blanket, we escaped to our neighborhood pool. My husband and I would take turns swimming laps in the bracing, deep-blue lap pool while the other one bobbed with the kids in the crowded, decidedly less refreshing family pool. One day, when my eldest daughter was about six or so, she wanted to try swimming laps. She wasn't a strong swimmer, so I sat down by the starting blocks to keep watch. She sliced a wobbly, meandering arc from one end of the pool to the other, each tiny arm rising in turn to catch the sunlight before it plunged back into the water: halting pinwheel of flesh and spray. With every flash of arm, I felt a lurch in my own brain as though, through sheer force of will, I might join my muscle to hers and so ease her perilous passage.

There is a seventeenth-century painting, *The Fall of Icarus* by Jacob Peter Gowy, that captures this phenomenon of shared psychic and muscular space binding parent and child. Icarus—having, as we know, flown too close to the sun—sprawls in the foreground of the canvas, tilted upside down. Wisps of unstuck feathers float lazily around his arms as if to mock the gravity that, even now, the viewer feels pulling at his body. Daedalus above him looks on in panic: foot flexed, knees braced against empty air, the muscles of his forearms rippled as he grips his wings more tightly. His entire body strains, as if with enough effort he might pull taut the helpless slack of his son's limbs cut loose in space. All to no avail; Icarus plunges earthward.

Yet the impulse I felt while watching my daughter swim—to trace the same arc with neuron and muscle, to mirror the path she cut across the water and the world—is less parental projection than a basic fact of embodied existence as a self among other selves in the world—perhaps *the* basic fact. In philosophy, the "problem of other minds" names the skeptical conundrum of how, trapped as I am inside my own head and its representations, I can be certain that other minds exist outside of my own. "By what evidence," John Stuart Mill mused in 1865, "do I know . . . that the walking and speaking figures which I see and hear, have sensations and thoughts, or, in other words, possess Minds?" Mill answered his own question by arguing from analogy: these walking and speaking figures have bodies "like me," the precondition of feelings and thoughts, and thus by analogy I can decipher the other's inner states by reading their body's "outward signs." The argument stuck, give or take a philosophical twist, throughout most of the twentieth century: the other is another me.

But in fact, the whole process by which the brain maps a self is much less unidirectional. Infants as young as twelve days old stick out their tongues, open and close their fists in imitation of gestures they see performed by their caretakers. The infant uses the other's face and hands as a mirror, psychologists argue, mapping its own body's boundaries by comparing "the sensory information from his own unseen motor behavior to a . . . representation of the visually perceived gesture [to] construct the match required."

The neural basis for such an intersubjective mirroring mechanism in primates was confirmed in a 1992 study that revealed the same neurons that fire when a macaque monkey reaches out to grab a peanut also fire when the monkey simply observes this action being executed by another. One researcher compares the neural mirror mechanism to the way predicates function in a grammar: a verb's meaning remains constant, independent of the agent performing the action—*I, you, they,* no matter—such that "the first- and third-person experience of a particular emotion [or action] are mapped by a shared neural representation." We create a body map not just in conversation with ourselves—with our own neuronal webs of visual, muscular, tactile feedback—but in exchange with other faces and bodies, whose external visual images we then match to our internal proprioceptive loops. It is only by slipping into the borrowed skins of these others—inhabiting their envelopes, internalizing their outlines—that I first become a self.

The other is, then, not "another" myself; it is rather closer to the truth to say that the self is another *other.* The distinction might sound like surface semantics, but I don't believe it is. It means the self is never whole to begin with, is always something of a knock-off. Pythagoras's monad without the dyad—the one without the two—is no-thing. This elemental fact of embodied existence, this secret neural overlap of first and third person perspectives, without which there would be no shared world at all, remains largely unconscious in our day-to-day functioning. It is only when we are cast into certain liminal states—like physically birthing another body, creating two out of one—that we sometimes catch a glimpse of the other we always already are.

When my daughter was six, she said she wanted to be a skeleton for Halloween. I bought a long-sleeved black leotard, thick black tights, and a packet of white felt and set to work. I cut out tibia and femur, radius and collarbone, spinal cord and pelvis, stitching each piece onto the black canvas: a constellation of tiny bones. They held a Halloween parade in the gym at her school, and a gaggle of witches and dinosaurs and robots marched in a circle while the parents elbowed each other and craned necks and held camcorders aloft to capture the moment.

I caught sight of my tiny parading skeleton, centered her in the camera's viewfinder, and zoomed in. In the enlarged image I was suddenly able to see—what hadn't been apparent with my naked eye—that her face was streaked with tears. Just then her gaze met mine behind the camera lens. Her face lit up with an expression of mixed longing and hope, and her skeleton arm shot out toward me as though, from across the gymnasium, the gesture might magically collapse the space separating us. With tears now streaming down my own cheeks, I snapped the picture.

TURNING, UNFOLDING, PASSING THROUGH

I went on a walk last March with my youngest daughter, picking our way around spring's effluvia laid out on the sidewalk like a carpet: the crushed top of a hummingbird egg; a black-and-red speckled insect I couldn't identify. "It's a ladybug larva, mom," she informed me. "Didn't they teach you that in school?" They hadn't, in fact. This creature looked like a ladybug rolled into a cylinder and pinched at both ends, a ladybug reflected in a funhouse mirror.

Larva derives from the Latin word for a ghost or evil spirit and came, by association, to name the masks used in Roman theater. In the seventeenth century, the budding science of zoology borrowed the term to describe the stage in an insect's life cycle where it is no longer egg but not yet adult: a ghost, or mask, of what it will eventually become. This thing was the mask of a ladybug, its distorted iteration; a flash of what, in the fullness of time, it would turn toward and into.

On other walks, we stumble upon some of the season's casualties: a dead fledgling sprung from its nest too soon. Bobbleheaded, violet eyelids and horny yellow beak, a scrim of pale blue feather-fuzz on its skull. Its body is plump and pink and wrinkled as an earthworm. When I pick it up it is doughy to the touch, light as a leaf.

We bury the baby bird underneath the nest from which it must have leapt; before its prime, before its time, this gangly ghost-shadow of its fuller future self.

Later that week we find an adult jay, resplendently blue and perfectly formed, lying on the sidewalk as if he has just fallen asleep there. My daughter wants to bury him. "I'm not running a bird cemetery," I object. She glares at me in silence until I go back behind the house and get the shovel. Now we have two miniature headstones nestled among the cacti in our front yard, each adorned with a painted heart the size of a fingernail: white for the fledgling, blue for the jay. My daughter scatters flowers on the gentle swell of each tiny grave.

The first real biologist, Aristotle, gathered evidence for his scientific treatises, *On the Parts of Animals* and *The Generation of Animals*, while living on the island of Lesbos. There, he studied specimens that he pulled from a lagoon: crabs and bony fish, squid and mollusks. He complained that his predecessors' efforts to catalogue the natural world had remained on the level of description, divvying up plants and animals into classes using such accidental characteristics as "configuration and colour." A true science, he insisted, must move beyond the surface, delve into process: "how each of these [animals] comes to be what it is, and in virtue of what force."

What counts in explaining any particular kind of life is not its substrate or matter—*hyle*—but its form or idea, *eidos*: the blueprint that determines what a thing must, of necessity, become. Aristotle uses the example of a piece of wood being sculpted on a lathe. "For it is not enough for [the sculptor] to say," he argues by analogy, "that by the stroke of his tool this part was formed into a concavity, that into a flat surface." Rather, "he must state the reasons why he struck his blow in such a way as to effect this, and what his final object was; namely, that the piece of wood should develop eventually into this or that shape." A sculpture is explained not by reference to its material (bronze or wood or clay), nor by the actions to which it is submitted (cutting or shaping). Rather, it springs from the blueprint (*eidos*) of the object that the sculptor seeks to craft, and that he holds in his mind's eye. Just so, an animal springs from the germ of its parent, which contains inscribed within it the *eidos* of the one and only animal it can become.

The temptation to read purposiveness into nature has dogged biology ever since. An American philosopher of biology, Edmund W. Sinnott, wrote of this temptation in a 1954 essay. No matter how many times we tell ourselves that adaptation (and thus evolution) is a matter of chance rather than intent, purposiveness continues to hover about natural phenomena. "To watch an animal egg cleaving in a precise fashion and marching inexorably on toward its culmination in the adult or to see a bud unfolding into an intricate system of leaf and flower . . . gives a vivid impression of activity that is goal-directed," he muses. "Whatever we may think of it in theory, the organism *looks* as if it was going somewhere." Or observe, for instance, how "the wing of a cicada is unpacked as the adult bursts its larval bonds," he notes. "No parachute was ever folded with the precision these organic structures show, the more remarkable since they develop from a tiny, formless mass of cells."

Many centuries after Aristotle floated his analogy between the purposive unfolding of an animal and the sculptor's art, the English verb "turn" evolved out of the Greek word for a turning lathe, *tornos*, parleyed into Latin *tornare*: the act of shaping, polishing, fashioning matter into form by spinning it on a lathe. Hence the English word's connotations of a process by which a thing—a "mass of cells," a hunk of wood, "without form, and void"—becomes what it was always intended to become: literally *turns into itself*.

Yet no form lasts forever. What do we call it once the *eidos* of a form is reached; once the thing a thing is supposed to become achieves peak thingness? We used to visit my husband's great aunt when she was in her eighties, meeting on Sundays to eat undercooked eggs and sugary bagels in a northeast Philadelphia diner. She kept her bottle-blond hair swirled in a fifties-era bouffant, a relic from youth that she carried forward with her. Her nails, shellacked in shiny burgundy, clicked against her glass of iced tea. But amid these reminders of an earlier iteration of herself—when she was young and colorful and naturally blond—what always transfixed me was her skin falling in soft bracelets around her wrists as she stood up.

As I age, I find myself physically turning into my mother. A certain way of inflecting my voice when I call my children pet names ("dear heart," I cluck, unbidden); these ever-thickening ankles. Occasionally I skip a generation and channel my father's mother instead: sometimes now when I eat, my left pinky finger levitates slightly, exactly as my grandmother's did as she chewed her breakfast toast.

These evolving tics and echoes do not go unnoticed. The other day my eldest daughter caught hold of the under-flesh of my arm in one of my unguarded moments and screeched, "It's soft just like Grandma Kate!" (Afterward she kissed the top of my head and said wistfully, "I don't want you to die.") And so it is: my skin is becoming soft as a dog's silk belly, soft with the slippery, ceding velvet of ripe fruit. "Past their prime," my mother says, shaking her head at pears forgotten on the kitchen counter, turning to rot.

My turning flesh haunts me when I chance to catch it in the mirror: the flap between neck and chin, thin and white as a skin on hot milk when you pinch it between thumb and forefinger. My speckled hands. Aging is a kind of anti-*eidos*, a downward tumble back into formless matter.

This past March, tiny brown painted lady butterflies descended on Los Angeles. The plentiful rain brought them winging their way north out of the Mexican desert to breeding grounds in Oregon. As I drove through the city, I mistook them for swirls of ash or paper. Clumps of them dipped and dove at my car, sucked into its undertow, before wind or a chance wingbeat jerked the lucky few up and away from the hurtling metal. The sheer volume of butterflies, and the hecatomb that I knew awaited them every day on the 405, was tough to stomach. Every day I wondered: was no species learning curve going to kick in to circumvent the carnage? But no, apparently not. Nature's potlatch marched on, cool, undeterred. The butterfly invasion lasted three days.

Nature's spectacular waste in pursuit of propagation is on display in the vegetable kingdom as well. Eye-catching fruits and flowers rig up baroque systems of seduction to get themselves reproduced. The bird-of-paradise plant's splay of crenellated orange and blue blooms join, at their base, to form a nectar pouch. When sunbirds perch on the stalk to sip, its petals coyly scatter pollen onto their feet.

Or take, for example, the tree *Hura crepitans*, informally known as the dynamite tree. Not content to stake its genetic fate on the whimsy and chance of passing animals, the dynamite tree's pumpkin-shaped fruits are capsules designed to crack open and literally explode, jettisoning seeds, like a spray of shrapnel, up to a hundred meters from the parent plant in a fearsome feat of vegetal ejaculation.

Perhaps none of us, from this evolutionary perspective, are anything more than highly complex seedpods; the fruit matters only as host such that, once its disseminating mission is complete, it may crumple up and rot.

Other times, I take the long view. Then biology seems less a straight line unfolding from birth to death, the inexorable wax and wane of form, than an endless shuffling and reshuffling of traits across time: we appear on earth, return to dust, only to reappear, in scattered bits and pieces, in our offspring ever after. The curve of a lip, the swing of an arm, the lilt in a voice: scraps of DNA chopped up and broadcast across the generations. My grandmother's nose is recapitulated, in diluted form, in the pinched flare of my eldest daughter's right nostril: a certain crookedness I, too, inherited and that once, at my grandmother's funeral, allowed me to pick out of a crowd a long-lost cousin.

I was at the refreshment table at the reception, after the interment at the cemetery, pouring a glass of sub-par white wine (the only kind my family has ever served at these functions) into a plastic cup when I glanced up to see a girl I'd never met before, and it was like catching a passing glimpse of myself in a mirror. I sidled up to a cousin who was picking at the fruit plate.

"Who is that?" I whispered.

"That's your aunt's daughter," she replied, nonplussed, popping a strawberry into her mouth. As though it were the most unremarkable thing in the world that my mother's younger sister, whom I had known for all twenty-six of my living years and who assuredly had no daughter, should suddenly be possessed of one. Growing up she was our favorite aunt, being more kid than adult. At Christmas she came bearing gifts from Boston: a bright green stuffed frog half my height; a lavender Indian pouch with tiny bits of mirror stitched into it; shiny red beans the size of a pea, hollowed out and filled with miniature white carved elephants you could shake out onto the palm of your hand.

But my aunt did have a daughter, it turned out: out of wedlock, born in a convent, given up for adoption decades earlier, a secret the entire family had never breathed word of until my grandmother had safely passed away.

Here, suddenly, against all chance, a second self dredged up from oblivion, unmistakable: my grandmother's cells had marked us both in precisely the same way. A shibboleth carved in flesh: *We belong to each other.*

Stranger still are the subterranean siftings of DNA across time, throwing off sparks that light up and then disappear, mutate and recombine, soften or stretch, lie silent only in order to flash up once more across the span of years to form improbable, heart-stopping constellations in the bodies and faces of your relatives. Sometimes you will only catch these ghosts in photographs, as if the blank mechanical recording surface of the camera could detect patterns of light, pick out secret angles and refractions, invisible to the unaided eye. My father had an older brother who died of polio when he was nine and my father was five. Among the photographs of Michael that survive is one taken on a summer day the year before he died, standing in front of his house. Shirtless and thin, baggy shorts cinched at the waist, one arm thrown across his chest with his fingers resting lightly on the opposite shoulder, he exudes the indolence of a body freed from the strictures of school clothes and cold weather, a body abandoning itself to the warmth of the sun. His eyes, ice blue even in this black-and-white image, fix the viewer fast: *here I am.*

As I scroll through Facebook, my eyes fall on a cousin's recent post, a professional portrait of her and her husband in the family portrait genre: a soft-focus background of indeterminate nature, a flash of green hinting at a garden. But what strikes me is the resemblance, around her teeth and mouth, and maybe the arch of the eyebrow, to Michael—the uncle I never had the chance to meet. The portrait has the effect of a visitation: Michael coming back, peeking out from beyond the grave for a just a fraction of a second. My cousin has unwittingly carried him all this way, a silent passenger still asking, shyly, to be remembered.

During the painted lady migration, my daughters and I went for a canyon hike. Because of the rain, the ridge we walked along was alive with blue-green scrub unusual for this normally rocky, desert landscape. The hills were blanketed in soft shoots and grasses, the color of dragonfly wings, tender as corn

silk. As the wind blew over and through them, they rippled in thrilling, zigzagging patterns, like velvet rubbed the wrong way then suddenly smoothed back down by an invisible cosmic finger. The painted ladies dodged and pitched in the wind, only occasionally stopping to rest for a wingbeat or two to warm themselves in the sun. Then they were gone.

These tiny creatures might make it to Oregon, I thought. *Or, they might not.* Only God knew for sure. Nature's harrowing disregard for the individual life-forms we hold dear—mother, daughter, father, son—has ever been a wound and a mystery. We imagine someone, somewhere, must be counting. And failing that, we may make a point of doing so ourselves: even the bobbleheaded fledgling found its way onto somebody's list—my daughter's—when she marked its passing with a tiny headstone. "Are not two sparrows sold for a farthing? And one of them shall not fall on the ground without your Father," promises the Gospel. "But the very hairs of your head are all numbered."

The lure of a final counting—eternity's roll call—has always been one of Christianity's greatest attractions. Who doesn't want to believe that what was lost will one day be restored? On the last day, the earth and the sea will give up their dead, all God's lost treasures rendered: then the dead "great and small" will stand before His throne, and "this corruptible shall have put on incorruption." The dream of resurrection is the dream of an *eidos* that stops turning, heaven putting earth's changefulness to rout.

We despair of matter—its inconstancy, its corruption, its fragility. Yet William James, who knew whereof he spoke (having lost his one-year-old son, Hermann), had this to say of matter: "To any one who has ever looked on the face of a dead child or parent the mere fact that matter could have taken for a time that precious form, ought to make matter sacred ever after." Form doesn't have to be eternal. The simple fact that matter accommodated *this* beloved form passing through ("for a time") is the joy and the wonder.

It is not everything, but maybe it is enough.

AMERICAN PASTORAL

In 1915 my great-grandfather Dennis Dwyer entered a church raffle. Against all odds, he found himself holding the winning ticket for a brand-new Ford Model T. Now that he had a ready means of transportation, he carried out a dream he had nursed since first arriving in the United States from Ireland: he bought a small plot of land on Lake Ontario, sixty miles northwest of his hometown of Syracuse, New York, and built himself a summer cottage. My grandmother and her sisters photographed themselves hanging on the car's running board in front of the green-shingled house, mugging for the camera in flapper scarves and drop-waisted shifts circa 1920.

Growing up in the 1940s and '50s, my mother spent her summers at that cottage; there is a photo of her at eight, maybe ten, years old, towheaded and knock-kneed, reclining in the white sand dunes leading down to the water. By the 1970s and '80s the dunes had disappeared, replaced by a shoal of slippery rocks along the lake's edge. But my sister and I could still scrape together enough sand to fill a bucket and spent hours sifting through the stones to find sea glass washed smooth by the surf.

Today, one hundred years after my great-grandfather hammered the last nail, the cottage still stands, if barely: it groans and rocks and creaks in the wind and damp, all buckled panes and warped wood floor, but it does not topple. Every summer when I turn off the dirt road and pull up into the front yard, my daughters and their cousins tumble out of the car and race to the water, wind catching at their hair. I watch as they pick their painful, barefooted way over the rocky beach, arms outstretched like tightrope walkers. In the afternoons we collect wildflowers, Queen Anne's lace and chicory and buttercups, and weave them into fairy garlands.

The fantasy of a retreat from civilization into nature, far from the madding crowd, is almost as old as humanity itself. As early as the eighth century BC, Hesiod was already lamenting the loss of a golden age when "the grain-giving earth bore crops of its own accord, much and unstinting," and men "lived entirely apart from toil and distress." Theocritus and Virgil wrote of shepherds and shepherdesses, nymphs and satyrs frolicking in the Sicilian sun. Inspired by these literary idylls, city-weary Roman nobles escaped from the "distress" and corruption of civilization by building the pastoral sensibility into the very architecture of their villas. Semicircular dining rooms opened out onto gardens so that residents could dine alfresco, just as Theocritus's shepherd made a bed of "fragrant reeds and fresh-cut vine-strippings" on which to take his simple country repast, "as many an aspen, many an elm bowed and rustled overhead." The most elaborate of these ornamental gardens contained streams, fountains, and concrete nymph grottos. Fast-forward over a thousand years, and we find the European leisure class still at it: in 1783 Marie Antoinette famously ordered a rustic hamlet built on the grounds of Versailles, stocked with live animals and real-life herdsmen and dairy maids, where she could dream of simpler times beneath thatch-roofed huts.

Yet my family cottage—fond as I am of it, and sylvan flower crowns notwithstanding—is not a place that lends itself easily to rustic reverie. Oswego County is thinly populated and among the poorest counties in New York State, with 19 percent of its inhabitants living below the poverty line. Driving from Oneida to Ramona Beach, as the interstate gives way to numbered county routes and then nameless dirt roads, we pass through hamlets with one intersection where the only show in town is a Dunkin' Donuts or a Taco Bell. We pass ramshackle farmhouses; collapsed barns; trailers with giant satellite dishes stuck like toadstools on top of them; abandoned country cemeteries, tipped and mossy headstones rearing up like a jaw full of crooked teeth. We pass by our Amish neighbor Ezra's farm, his cornflower-blue-and-black-clad sons behind a horse-drawn plough in the July heat. Land is cheap here, bringing an ever-growing community of Amish. Further on, in front of one paint-blistered house we see a hand-painted sign planted in the grass: "SWEET CORN AND AMMO FOR SALE."

Our cottage is nestled on the edge of Mexico Bay, a shallow crescent on the Lake Ontario coastline. In 1969, the year I was three and my sister was born, the Nine Mile Point nuclear power plant was built in the county seat of Oswego, twenty-five miles down the coast. It was the height of the Cold War and the Vietnam War. Atom splitting, for more than twenty years associated with the horrors of nuclear holocaust and radiation sickness, was attempting to rebrand itself as the energy of the future. New York Governor Nelson Rockefeller was in Oswego for the Nine Mile Point dedication ceremony, touting nuclear energy as a giant step for humanity in our ongoing quest to bring light to the darkest recesses of the universe. "On an occasion such as this, it hardly seems possible that a brief seventy years ago, electrical power was a curiosity rather than a necessity," he observed. "Today, when the power stops, modern living stops. We get an instant reproduction of the Dark Ages." He knew all about the perils of splitting the atom, he assured the audience. But "we live in the nuclear age, and we can no more turn back the clock than could our ancestors in the age of steam or when fire was first discovered." The sensible course was not to fear nature's inscrutable power, but to "understand it, to master it and turn it to our advantage."

So now, as my sister and mother and I sit in our aluminum chairs on our shrunken patch of beach, sipping our morning coffee, we spy the twin nuclear reactors at the far tip of the bay: squat and gray, giant white plumes of water vapor melting into blue sky. Nine Mile Point has the dubious distinction of being the oldest functioning nuclear reactor in the country. We joke about three-eyed fish, radioactive seaweed. How cousin Roxane never let her kids get near the water.

The dream of a pristine state of nature has always been a ruse—innocence recollected from the rueful perspective of experience, which is to say, not innocent at all. This, certainly, is no arcadia, no untouched idyll: for every pound of sweet corn, a pound of ammo; for every flower crown, a sluice of radioactive waste. The marks of human artifice, here, hit you like a poke in the eye.

Human-induced blight isn't the only sinister element in the landscape around Ramona Beach. Nature, too, is menacing, imposes its own mysterious scourges. There were several summers during my childhood when heaps of dead alewives (we called them mooneyes) would inexplicably wash up on shore. "Few things look as dead as a dead alewife," observed one science writer of these mass die-offs in the 1990s, and a truer sentence was never written. We would wake to find the mooneyes littering our beach front, as if the lake were a housecat faithfully depositing its daily kill on the owner's stoop. Their perfectly round, silver eyes, pitted with giant black irises as deep as wells, lent the mooneyes an air of shocked indignance, as though death had caught them by surprise. They rocked lazily in the surf, fixing the sky with sightless orbs. By midday they reeked so much that my father had to shovel them into buckets and haul them away.

Other times it was algae: thick, pea green, an invasive vegetable mucous. We went swimming anyway and emerged from the depths dripping like shaggy green Neptunes, soft globs of plant sticking to our suits, our skin, our hair.

Then, a few summers ago, Ramona Beach was visited by a plague of frogs. They were suddenly everywhere: hopping down the oil-slicked dirt road, rustling in roadside weeds, squatting quietly camouflaged in tufts of grass on the front lawn. Their skin was exquisite, luminescent silk swirled with a paisley of mud brown and peacock green. The frogs were so plentiful that after a day the kids didn't even bother to catch them. A captive frog thrills only by reason of its rarity, cupped hands slivering open to reveal a pulsing emerald gem against pink flesh: nature neatly imprisoned for our scientific or aesthetic pleasure. But outnumbered by the frogs, it was we who began to feel like interlopers in this landscape.

Far from wishing to catch them, we grew weary of the constant vigilance it took to dodge and shield ourselves from them, to avoid treading them underfoot and turning them into hash. One day my daughter was jogging down to the creek in flip-flops when a frog with an unfortunate sense of timing leapt, lodged itself between her heel and sandal midstep. She limped home and we had to scrape bits of smashed frog off the sole of her foot.

Welter of frogs, raft of mooneyes, glut of algae. Nature's unbidden, often ruinous, surfeits have always been cause for human soul-searching: Whose wrath have we piqued? Which god have we slighted? Let my people go, called Moses; Pharaoh hardened his heart, and we know the rest: frogs, lice, flies, boils, hail, locusts rained down on Egypt. Nature's whimsy flouts the best laid of human plans, and it is no wonder humans had to imagine a murderous divinity to account for these assaults. Someone who would listen to reason or—failing that—would listen to burnt offerings and prayers, grant forgiveness for our trespasses, restore nature's disordered rhythms.

But if in biblical times the wrath of God's nature was largely mysterious, punishment for secret sins, today we know all too well whence these ever-escalating swells and gales and gluts and droughts come. We fret; we eye the algae with growing panic. What infernal machinery has raised the water temperature by a hair's breadth, killed off the algae-eaters, unleashed this ghastly green tide? We count the frogs, chat with neighbors. Could their population rebound possibly be a good sign—nature righting itself? Or is it, after all, another click of the climate change wheel as it whirls off its axis?

We have sickened the earth, and now it is coming for us.

A few hundred yards down the dirt road from our cabin sits Snake Creek, a stagnant basin where water from the seventy acres of marshland and forest that fan out behind it comes to a standstill before trickling out into the lake. The creek is brown as molasses and almost as thick, edged with cattails and pickerelweed and jewelweed. Jewelweed plants pump their bright green seedpods full of water so that the casing is stretched to bursting; at the lightest touch the pods explode, sending seeds flying. These sudden firings make the plant stems arc and crackle, as if the whole bank of vegetation were alive. As kids we wandered back along the creek edge popping jewelweed and picking horsetails until the ground became spongy and then we knew we were no longer on terra firma but had crossed over into swamp. We'd scuttle back to the safety of the road like insects fleeing a spider web.

Snake Creek is home to snakes, turtles, frogs, leeches, snails, and all manner of freshwater pond-dwelling fish from the cyprinid family—carp, minnow, sunfish. Cyprinids are a primitive class of fish with toothless jaws and no stomachs; they suck at their food, mash it between jawbone and palate like babies gumming applesauce. Swamps provide vast, murky mounds of the decaying biomatter, plant and animal, on which they feed. We jerry-rigged fishing poles and went out early mornings to the creek, using Oreos for bait. Now and again a giant carp would surface in the distance, mouth gaping, then disappear among the cattails. While the big fish ignored our cookie-crumble bait, we caught several small sunfish this way, lured to the edge of the sunless lagoon by the taste of industrial-grade corn syrup.

I always got goosebumps sitting there in the quiet, looking out across the mist-swirled swamp, slip of a sunfish dangling from my pole. In that marshy, fathomless netherworld, I knew humans didn't stand a chance. In an instant the world could flip: catcher become the caught, fisher become the prey. That swamps (marshes, quicksand, lagoons) pose a threat to individual human life and to civilization as a whole is a locus classicus of American horror fiction. Swamps always cue the possibility of sudden, brutal payback for mankind's transgressions.

Governor Rockefeller's paean to "the peaceful power of the atom" notwithstanding, some of the most iconic American monster movies ever made emerged in response to the specter of atomic energy. In the 1954 cult classic *Creature from the Black Lagoon*, a paleontologist in the Amazon uncovers the fossilized remains of a giant webbed claw. He whisks it to the nearest science lab where they surmise that it may be one of nature's first, abortive attempts to catapult life up out of the seas and onto land. They launch an expedition to the Black Lagoon, in the darkest heart of the jungle, to search for the rest of the skeleton.

What they find instead is "gill-man," a Devonian-era human-reptile hybrid, a missing link who is still alive and well and out for blood. Despite their fancy boats and winches and diving equipment and evolved brains, humans meet their match in gill-man, who is supernaturally strong and has home-court advantage in the swamp. Scores of brown natives (too childlike) and white scientists (too soft, effeminate) meet grim watery deaths before the team finally devises a muscular, brainy way to eliminate the creature. As the closing credits roll, gill-man slowly sinks back to the bottom of the lagoon where he belongs.

Other atomic-era films were more pointed still in their efforts to reassure Americans that humans could mine and exploit nature without fear of any real reprisal. The year 1953 saw the release of *The Beast from 20,000 Fathoms*, which opens as a posse of cheerful scientists are carrying out nuclear tests in Antarctica. "You know, every time one of these things goes off I feel as if we were writing the first chapter of a new Genesis," gloats one technician as he watches the mushroom cloud blossom. The hero of the story knows nature is trickier than that. "Let's hope we don't find ourselves writing the last chapter of the old one," he warns. Man has no inkling what the cumulative effects of all these atomic explosions will be; the world has been here for billions of years, while "man's been walking upright for a comparatively short time," the wise physicist cautions.

Lo and behold, the blast dislodges a Mesozoic era lizard trapped beneath the ice. Following an ancient homing instinct, the beast swims down the Atlantic coast from Newfoundland to Maine to Boston until he reaches home: the subterranean canyons in the Hudson River off Manhattan. He clambers ashore just north of Wall Street, crushing cars and munching policemen as he makes his way up to Times Square.

In the end, it is only by taking the power of atom-splitting and turning it against him that the monster's destruction is assured and man's absolute right to manipulate nature as he pleases reaffirmed. The monster's blood is filled with a dangerous pathogen; only by neutralizing the microorganism with the curative properties of radioactive isotopes—curing and *then* killing the beast—can the threat of human annihilation be avoided. This is a job that requires both brain and brawn: a scientist to provide the radioactive material and a trained solider to shoot it into the animal. "Do you know what a radioactive isotope is?" the nuclear physicist asks the soldier. "No, but if I can load it, I can shoot it," he tosses back. Donning hazmat suits, the two men set off to slay the dragon who has now run amok in Coney Island. Yes, splitting the atom destroys cities and unleashes prehistoric hell. But it also cures diseased tissue, so it can't be all bad.

Last summer at Ramona Beach we made friends with a dusty old dog, a hefty German shepherd–Saint Bernard mix, who licked our hands and followed us back to the cabin where we fed him scraps of lunch meat. My eight-year-old nieces named him Seymour, but we eventually discovered his real name was Cujo—after the rabid killer dog in Stephen King's 1983 horror film of the same name. In the movie, Cujo is on a forest frolic chasing rabbits when he stumbles into a cave and is bitten by a bat. But it is not clear whether nature or man is the true culprit in the demonic transformation that ensues. Cujo's owner is a feckless, poor, wife-beating rural car mechanic. When Cujo rips his throat out atop a trash pile of rusty bed springs and empty beer cans, surely he had it coming.

Nor, in the movie, are the affluent spared the vengeance of nature outraged. Across town lives a wealthy advertising executive from the city. He plays tennis and drives a red jaguar convertible, but the raspberry breakfast cereal he markets has just been recalled for making kids pee and puke red dye. "Nothing wrong here," runs the cheerful jingle shilling poison cereal on the television. But we suspect otherwise: nature has been violated, not only by the shiftless poor but by greedy corporate types as well. Payback will be swift.

When the ad exec's son and wife are inadvertently caught up in the Cujo mayhem, it comes as no surprise.

The film doubles as a prescient '80s-era class parable, one that, foreseeing the escalating ravages of extractive capitalism in America, nevertheless scrambles to reassure the audience that we will all be OK. Cujo could only be spawned in the derelict backyard of a poor white-trash mechanic, and only the ad executive's rich wife—chastened, to be sure, but still rich—can ultimately reestablish order. We didn't know it yet in 1983, but the rural poor were on their way to getting much poorer, while the corporate wealthy were poised to consolidate themselves as the unreachable 1 percent. *Cujo* hedges its bets, acknowledging evil on both ends of the economic spectrum, but ultimately places its faith in a middle term—here represented by a white housewife, able to wield a gun when necessary, but only in defense of maternal values—who can bridge the gap and restore both natural and socioeconomic balance to the heartland. That balance *will* be restored in the end is never up for question.

I have to squint to see the resemblance between the dog in front of me and movie Cujo: nothing could be less threatening than this fat, shambling, doe-eyed creature angling for tummy-rubs at our feet. Then again, who can be sure? The domesticated is always capable of rewilding; the civilized always risks sliding back into the savage. I side-eye this possible turncoat in our midst.

Horror movies know, of course, that nature's monsters are only ever a proxy for the deeper threat of darkness coiled like a serpent in the human heart. "It's not a monster, it's just a doggie!" screams the ad exec's wife to her terrified child, all evidence to the contrary: the bloody, slobbering beast is lunging at them through the cracked window of their Ford Pinto. That's the fairy tale we tell ourselves, tell our children, and one with which I am all too familiar. No, it's not climate change; this is just regular summer, a little hotter maybe. No, it's not end-times. Humans are resourceful: we'll find a way, we'll stop global warming—don't you worry. We'll shoot it full of radioactive isotope.

For the moment, gathered around Cujo on the front lawn, we feel trusting. It is so much easier to believe that it all turns out alright in the end. That one day we'll learn not to waste the earth and the souls of the people who labor on it. We hand-feed snacks to the shaggy beast; we ruffle his oily fur and knuckle his head. A temporary truce.

"I'm done," my sister declared one summer in exasperation. The lake algae that year was particularly thick, impossible to rinse out of the twins' knotted hair, and the outdoor shower was filled with spiders. It was the same summer that, pulling up our beach chairs to watch the sunset, we kept spotting water snakes: a few feet from shore, their heads held just above the surface of the water, jerking their way like mini Loch Ness Monsters back to the mouth of the creek. Were they water moccasins, we worried? Would they bite our children? The nuclear reactors glowered orange in the dimming light.

But of course, she wasn't done, and neither was I. We kept coming back, and our children would eventually grow to love-hate the place with the same degree of intensity. All we have is this fallen planet, and fallen things require our love—even when we suspect that love isn't strong enough to save them, or us.

In the twilight I watch as my younger daughter pedals off down the dirt road, executing a series of lazy dips and turns atop a rusty bicycle she salvaged from the neighbors' trash. This human-made machine left for scrap still holds together, barely. Above her, the canopy of blackening trees above the swamp pulses with hundreds of fireflies, winking like strings of white Christmas lights, as if the stars had been netted and dragged down to earth, an artificial firmament. I watch and listen as the creak of bicycle springs grows dimmer and her silhouette is swallowed by the dusk.

THIS RAGGED CLAW

One day that fall I come home from work to find, among the usual pile of bills and fliers and mass mailings pulled from the mailbox, a crisp white letter from the hospital where I'd recently had a mammogram. I am used to these letters, which appear every year, a week or two after my annual appointment, confirming my test results as normal and reminding me to schedule another mammogram in a year.

This year the form letter deviates from script. "Your test results were incomplete," I read aloud. "We recommend you come back in for further imaging."

My husband looks up from the dining room table where he's working. "Is this something we should worry about?" The modern English sense of the word "worry"—to feel "mental distress or agitation"—is of relatively recent origin. The original Old English word was much bloodier: *wyrgen*, to "strangle," or to kill by biting and shaking an animal by the throat. It is the wolf going for the jugular, shaking the lifeblood out of its quarry; the dog incessantly worrying his bone. Other English words denoting mental pain or obsession have similar roots in animal acts of eating or attacking: to *fret*, cognate of the German *fressen*, "to eat up." There is also *gnaw* and, relatedly, *gnash* from Old English *gnagen*. It comes to us via the proto-German verb *gh(e)n*: onomatopoeia for the act of gobbling up by little bites. We are consumed by grief; gnawed by remorse; eaten up with worry.

I glance at the diagnostic code at the bottom of my mammogram letter—"BI-RADS 0"—and quickly google it. "A score of 0 indicates the images may have been unclear or difficult to interpret," I learn. "Often women 40 years and older receive scores ranging from 0 to 2, indicating normal results or that abnormal results are benign or noncancerous."

"No," I reassure my husband. "No need to worry."

Still, I dutifully return for a second test. After the mammogram, the nurse takes me to a room where I'll have an ultrasound. I've complimented the staff on the upgrades they've made since my last visit. In the waiting room I find iced cucumber water in a glass dispenser, and in the examining room, fluffy terry robes in place of the usual pale blue paper slips. There is even a bowl of fruit—apples and oranges—cheerily posted on a table outside the changing stalls. "I feel like I'm at a spa," I quip to no one in particular.

Prone on the examining table, I slip my left arm out of the robe and feel the warm silicone gel slide over my breast and chest. Then the wand, swiping back and forth against my skin. The image on the screen, grainy black and white and gray, could be a picture of anything: riverbed, water rushing over rocks; the smear of a galaxy; the salty sea of a womb quickening with life.

As it happens, it is none of the above, but a clump of cells the radiologist doesn't like the look of. A white smudge nestled against the striated black soup. It is round and smooth on one end, but juts out with a jagged claw on the other. "What is it?" I ask the doctor.

"You see, a cyst or a benign mass would be perfectly round—like an egg or a pea," she begins. "This ragged edge"—pointing to the claw—"that's what makes it suspicious."

"Is it cancer?" I ask. It is only now occurring to me that it might be.

"I'm so very sorry," she says.

Cancer has always been imagined as a biting, grasping, greedy beast. It was Hippocrates who first named the disease *karkinos*, or crab, as "its veins are filled and stretched around like the feet of the animal called crab." It was an image that would stick, more deeply and vividly embellished by physicians ever after, as historian Alanna Skuse shows in her work on cancer's evolving zoomorphic identities. Like the crab, cancer was tenacious. It is "verie hardly pulled awaie from those members, which it doth lay holde on, as the sea crabbe doth," remarked one sixteenth century physician. There was no use in cutting away the tumor, despaired another observer, just as there was no forcing a "Crab to quit what he has grasped betwixt his griping Claws." Cancer the disease was as sneaky as its namesake. "It creeps little and little," noted one medieval commentator, "gnawing and fretting flesh and sinews slowish to the sight as it were a crab."

This gnawing behavior led early physicians to compare cancer to a worm as well as to a crab. The medieval name for the plant-devouring green caterpillar, *cankerworm*, derived from one such metaphorical leap from biology to botany: as cankers on the skin, so cankers in the bud. And just as malignant larvae in plants had to be destroyed before they blighted the flower, so one had to "slea the worme" of cancer when it chanced to rear its head in human flesh. This worm could be quite literal; the seventeenth-century surgeon Pierre Dionis surmised that cancer was nothing more than a "prodigious Multitude of small worms" infesting its host. A common early modern remedy, the so-called "meat cure," involved laying slabs of fresh chicken or veal flesh on the ulcer, by which to lure the creature out. Could the cankerworm be convinced to ingest the decoy flesh, the patient might be spared.

One tale that found its way into an early eighteenth-century medical manual—acknowledged by most even then to be apocryphal—told of a cancerous ulcer inhabited not by worms, but by a wolf. A witness vowed that in applying a slice of meat to a patient's tumor, "the Wolf peeps out, discovering his Head, and gaping to receive it." The existence of a literal cancer-wolf within the body may have strained belief. Still, the analogical force of these animals was strong: both the worm and the wolf made their way into the apothecary's shop as treatments for cancer. Following the homeopathic principle that "like cures like," one early recipe for a cancer-fighting salve included among its ingredients powdered "wolf tongue." More common were tinctures that called for ground and powdered worms. Worms harvested from tree bark, then "stamped and strained with Ale," were brewed into a potion eagerly quaffed by cancer sufferers.

Early modern physicians were not entirely wrong in imagining cancer as an eating disease. As the cancer grows, it needs more food. The cancer cells release chemical signals to the surrounding capillaries and veins coaxing them to create new blood vessels, to branch out like a coral forest, delivering life-giving nutrients and oxygen to their blooming colony. The body obliges, unwittingly nourishing the assassin nestled in its heart.

Hippocrates's *karkinos* would extend its spindly legs outward through the centuries, its metaphors spreading like creeper vines across the pages of the medical texts that sought to describe and contain it. Cancer is a crab; no, it is a worm; no, it is a wolf. As if the symbol that originally stood in for the disease could not help but mirror the proliferating, perambulating logic of the very thing it named. Words (cells) gone dizzy with division, malignant blooms without borders.

After the mastectomy, after the biopsy, the surgeon tells me she excised the one visible tumor but that the tissue extracted still showed a "positive margin," meaning they discovered additional cancer cells—ones that hadn't shown up on the mammogram—at the edge of the cut tissue. "The margin is described as negative or clean when the pathologist finds no cancer cells at the edge of the tissue, suggesting that all of the cancer has been removed," says the National Cancer Institute's dictionary of cancer terms. "The margin is described as positive or involved when the pathologist finds cancer cells at the edge of the tissue, suggesting that all of the cancer has not been removed."

The scalpel is clean and gleaming and precise, everything the cancer is not. It will slice through and extract the cells, like a paring knife scooping a bruise out of a fruit. But my cancer has evaded both the X-ray and the knife, the eye and the blade: a slip of a fish sliding off a hook, a mercury bead scattering at the touch of a finger. Now the fugitive cells could be anywhere, or nowhere. They elude detection. Absence of ocular proof is no reason at all to believe the cancer isn't there; on the contrary, absence multiplies the paranoid suspicion of its lurking presence, as Othello well knew. The surgeon sends my cancer cells back to the genetics lab to measure the chance of recurrence.

So, we will bring in the big guns, my doctor tells me, the chemicals and the lasers, to roust this worm from its secret perch. "O Rose thou art sick," writes William Blake in *Songs of Innocence and Experience.* "The invisible worm, / That flies in the night / . . . Has found out thy bed / Of crimson joy: / And his dark secret love / Does thy life destroy." I imagine the maniac cells coursing through my veins, caustic threshing of rosy flesh, seeking out the crimson bed of my liver, my bone marrow, my brain—plotting their return.

The summer we went to Crete, my daughters and I gathered shells at the sea's edge. They were beautiful, flashes of white against a bank of shiny dark pebbles. We found miniature conch shells, pulsing orange and pink; lavender sea urchins the size of a baby's fingernail. The spongy inhabitants had long since melted back into the salt water whence they came, leaving behind these bony memorials. When we got home and spread the shells out on the grass, one of them began to shake, made a hairsbreadth lurch, started to crawl: not mineral but animal, not husk but home. A tiny hermit crab had donned a stranger's skin, a twice-born shell. Its ragged claws peaked out and immediately retreated upon the lightest touch.

The next day when we woke up, the shell was still: its inhabitant tucked up inside, its fortress become a tiny tomb. My daughter wanted to give him a sea burial. She built a miniature raft of driftwood, bound together with some old string we found in the house we were renting. She wrapped the dead crab in a leaf and lodged him in a crook of the boat.

That night after dinner in town we walked to the beach. My husband lit the funeral pyre with a match and, cupping the flame with his hand, waded knee-deep into the wine-dark sea, placing it in a trough between two cresting waves.

No sooner had he let go than a foam-flecked swell swallowed the bark whole, downed it in one unceremonious gulp. From the shore, we couldn't help but laugh at nature's indifference to our tender rite. A message from the watery waste before us: no, your creature-kindled warmth is no match for me.

Cancer forces you to recalibrate your sense of danger and the direction from which it might come. Cancer is not the wolf at the door, begging entry of the three little pigs. No; this sickness is an inside job.

Kafka's story "The Burrow" is told from the perspective of a forest animal who has crafted a perfect underground hideout to protect him from predators. His is a world of frantic quick strikes, snatching jaws, eat-or-be-eaten. He muses on the less fortunate animals above ground, "poor homeless wanderers in the roads and woods, creeping for warmth into a heap of leaves or a herd of their comrades, delivered to all the perils of heaven and earth!" Tucked away in his burrow, he is safe; all is still.

Then one night he is awakened by a small whistling noise coming from within the walls of his home, almost nothing, "audible only to the ear of the householder." He scratches in the direction of the whistling, rifling through the soil, breaking up the lumps into tiny particles—"but the noisemakers are not among them." He steels himself; he will be methodical. "I shall dig a wide and carefully constructed trench in the direction of the noise and not cease from digging until . . . I find the real cause," he determines. "Then I shall eradicate it, if that is within my power, and if it is not, at least I shall know the truth. The truth will bring me either peace or despair, but whether the one or the other, it will be beyond question." But such clear-cut knowledge is not to be his. The noise seems to come from nowhere and everywhere at once. No matter where he digs, the noise persists: patient, steady, undeterred. It neither advances nor recedes. This is a threat that refuses to give him a toehold, a purchase, a vantage from which to confront it.

The outside world is a noisy, chancy, clanging place, chock-full of danger. This much I acknowledge. As the burrower observes, "danger lies in ambush as before above the moss." But from now on I know, like Kafka's animal, I will always be listening instead to that ghost of a whisper coming from inside my own walls.

Still, most days I don't worry. To be worried to death is, quite literally, the fate of all flesh. The fretting agent may differ, but we all know how it ends: our animal cells, the sick and the well, will sooner or later dissolve back into earth, be eaten up in their turn. So, I bend my ear, not to the inner rustlings of my cells, but to the small wonders of the world above the moss.

When I'm feeling well, I swim outside in the fading Los Angeles dusk. The pool is warm, 85 degrees to the ambient January 60-something of this desert climate. One night after I had finished my laps and popped my head up above the blue surface of the water, I heard clutches of birds singing in the surrounding bushes. A low chuck-chuck-chuck; a three-note trill; an impossibly dulcet single note that punctured the air like a plucked harp string, rising above the chatter. This wasn't a settling-down-to-sleep song. It was a raucous, drunken chorus, like the birds had forgotten what time it was, pouring their silken throats out into the gathering darkness. I crouched in the water, still except for my breath, and listened.

To accept our animal self means to look worry clear in the eye, to size it up, and then choose, instead, to wait. *To wait* is the most anodyne of verbs, a familiar no-man's-land of lost or useless time that threads itself through daily life: we wait in line at the pharmacy; we wait for a thundershower to pass; we wait for a friend to call. This waiting is empty time, a lull in the forward rush of life, of *getting things done*. It stops us in our tracks and temporarily suspends our power. It can make us feel annoyed or helpless. "There's nothing to do but wait," we tell ourselves. Then the prescription is delivered; the sky clears; the phone rings—and off we go again, our power *to do* and *to be* mercifully restored.

But there is a larger waiting, a cosmic waiting that precedes and cradles within itself all the other times and modes and tenses of being, like drops of water in an ocean. It tells us that our puny power to do and to be in this world is the exception, not the rule; that waiting is not the suspension of human business-as-usual, but rather the oldest and most elemental form of time.

The modern English verb *to wait* dervies from a trio of Old English roots indicating states of watchfulness and wakefulness: *waeccan*, "to keep watch"; *wacian*, "to be awake," and *wacan*, "to become awake, arise, be born." This is waiting as sheer presence, watchfulness, the quickening spark of life. It is *Ruach Elohim*, the breath of God, moving over the face of the waters the instant before creation: time before time.

To wait in this way means simply to bear witness to breath, the gift of wakefulness, as it lights up (here, now) in this accidental vessel that is myself. It asks nothing more of the time that remains.

NATURAL MAGIC

To have cancer is to take an unasked-for trip inside the body, to shine an unwelcome light into the body's normally invisible workings. It is to see the self (or a part of the self) lit up on an ultrasound screen, a white mass nestled in the grainy gray flesh of your breast, and know—not in a passing way, but in a fateful, here-to-stay way—*that's me.* And then you dive in even deeper: you understand that the white mass is there because there is disorder in your house at the cellular level.

Like rude party guests, your cells have ceased to behave: they do not wait for the proper signal from the host to invite more friends, but multiply with raucous abandon, overrunning the house. Normal cells are governed by a principle called "contact inhibition": they sense the presence of other cells in the room, and when the room gets too crowded, they politely stop dividing. Cancer cells don't respond to these chemical cues, and fill the rooms to bursting, spilling out the windows, passing out drunk in the backyard.

But to reckon with the fact that your life quite literally depends upon this microscopic dollop of protoplasm and its antics is to see life in a whole new way. The more you think about the complex mechanics of cells and the body, the trillions of tiny signals and movements and chemical shufflings that undergird a single second of a living body's existence, the more improbable your own existence seems. It is a cliché—one that is nonetheless true—that grave illness and brushes with death heighten our sensitivity to the numinous, bringing us closer to God, or at the very least to a chastened understanding of "what really matters" in life. This is because illness is a radically decentering experience, revealing the self as dependent upon some larger unseen thing that humbles by its sheer complexity and magnitude: call it biological life, the earth's ecosystem, the cosmos, God.

To think of what must happen, at a cellular level, to allow me each morning to twist my hair into a bun, apply a shade of age-appropriate lipstick (Revlon, "Nocturnal Rose"), grab my keys and coffee and head out the door—to glimpse the vast, mysterious architecture that enables and gives its blessing to that "self"—is to recognize the almost comical contingency of any single life. It is to realize how little in possession of our lives we actually are, how little they belong to us.

We secular moderns are ill-equipped to deal with this decentering, to make sense of our individual lives as embedded in a larger whole (whether biological or theological). But for millennia, the "self" was unintelligible outside of its participation in the cosmos. From ancient Greece through early modern Western thought, the human body was imagined as a microcosm, replicating in miniature the structure of the universe. No motion of the stars or planets was without its corollary motion in the world of earthly creatures below.

Before cancer was a disease, it was a constellation: one of a dozen-plus star patterns Babylonian astronomers identified as early as the second millennium BC, and whose movements across the night sky they used to divine the future. The Greeks gave the Babylonian crab a backstory: as Hercules was fighting off the Hydra in one of his twelve labors, Hera sent a giant crab to wrestle him and slow his progress. Hercules, undaunted, crushed the creature handily with his heel. Hera collected the shards of the crab's body and placed them in the night sky, a reward for the poor beast's loyalty.

By the first century AD, the Alexandrine astronomer Ptolemy had codified the predictive art of star reading in his *Tetrabiblos*, or *Quadripartite: Being Four Books of the Influence of the Stars*. Ptolemy described the universe as a series of ten revolving nested spheres, containing the sun, moon, planets, and the stars of the zodiac, with the earth as its fixed center. The earth itself and everything upon it, including the human body, was composed of four elements—earth, water, fire, and air—whose order and balance mirrored the motions of the heavens. Because human and animal and plant and mineral bodies were created from the same material as astral bodies, they necessarily exerted an influence on one another, a web of secret sympathies and correspondences that bound the universe together.

That the sun's motions should influence sublunary bodies was evident enough. But the positioning of the moon, planets, and stars was of equal importance in deciding the fate of earth's creatures, and "all the [heavenly bodies'] various influences compounded together," once expertly measured and calculated by the astrologer, could help predict "the destiny and disposition of every human being." This was especially true in matters of health, as the four humors composing the human body—blood, phlegm, black bile, and yellow bile—rose and fell, like the ocean's tides, in response to the distant tug of celestial motions. When the humors in a body became unbalanced, blocked, or thickened, sickness ensued.

Early modern physicians consulted astronomical tables when treating patients. Medical almanacs contained charts for calculating lunar cycles as well as tracking the hourly position of the moon, sun, and other planets as they cycled through the signs of the zodiac. When the moon was in its first and third stages, the body's liquids were thought to cluster at the surface of the body like tidal waters, making it a propitious time for bloodletting to drain off polluting humors. To help with diagnosis, the medical almanacs frequently included a jewel-colored anatomical illustration of the human body, homo signorum, nicknamed by historians Zodiac Man.

Zodiac Man stares out impassively at the viewer, a map of the heavens superimposed on his outstretched body. The figure stands atop two fish representing Pisces, while Leo the lion fills the heart and chest cavity, and a muscled Taurus sits coiled around his shoulders: to each body part, its corresponding star cluster. The moon was thought, like a cosmic magnifying glass, to intensify the power of each zodiac sign so that the brief moment of their crossing each month was sure to signal a critical phase in any associated ailment. The careful practitioner who sought to cure a patient of headaches

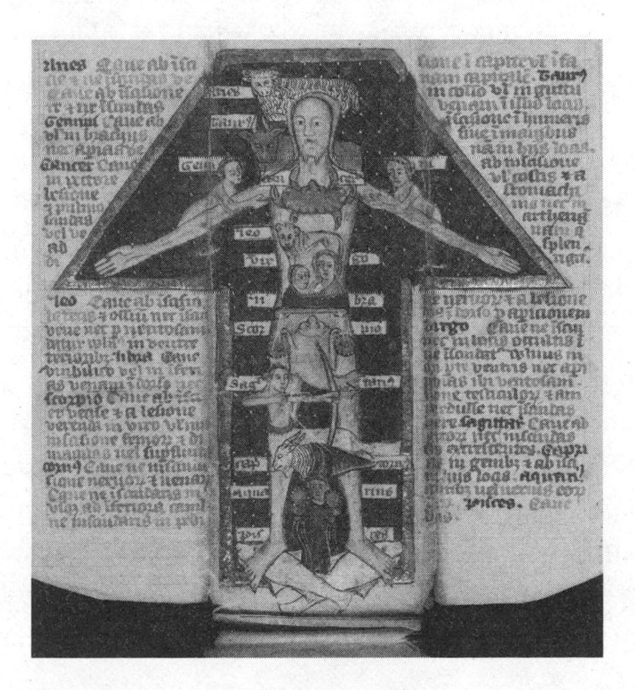

knew, for example, not to undertake bloodletting when the moon was in Aries, the constellation responsible for sicknesses of the head.

Once the blood was drawn, the physician carefully examined it for texture, color, smell, and even taste, as a means to further gauge the nature of a patient's illness. Greasy blood indicated that the liver was failing to concoct ingested food properly, and an herbal elixir—gathered and compounded when the heavens were propitiously aligned—might help to right the organ's function. Robert Turner's 1664 plant encyclopedia lists English native plants and their healing virtues alphabetically. White maudlyn, a sweet garden herb under the sway of fiery Jupiter, helped stimulate "cold and weak livers." Water Betony, ruled by cold Saturn, acted as a "good cooler in Burnings and Scaldings," and a paste made of sheep's dung, hog lard, and leaves of betony smashed in a mortar was "likewise good to dissolve swellings and hard knobs."

Marsilio Ficino's 1489 *Three Books on Life* gives useful tips on how to augment the body's vitality through medical astrology. "If you want your body . . . to receive power from some member of the cosmos, say from the Sun," he counsels, "seek the things which above all are most Solar among metals and gems, still more among plants, and more yet among animals, especially human beings." Among the created things with which contact could boost the solar composition of the body: gold, chrysolite, amber, saffron, aloewood, hawks, swans, and people with blond hair. "The above-mentioned things can be adapted partly to foods, partly to ointments and fumigations, partly to usages and habits," he advises. If you think your liver is failing to do its appointed work of heating the ingested contents of your belly, "draw the power of the liver to the belly both by rubbing and by fomentations made from things which agree with the liver," such as chicory, endive, and pastes concocted from animal livers.

So it was that nothing in the created universe existed in and of and for itself; leaf and stone, bird and beast, planet and human were not strangers, one to the other, but secret-sharers, speakers of a common cosmic tongue. So it was that the physician, at his patient's side, could taste the stars in a drop of blood.

The English word *zodiac* is derived from the Greek *zoidiakos kyklos*, meaning "circle of little animals." Early modern illustrations of the zodiac look like aerial views of a carousel, a menagerie of sign-creatures—lion, bull, scorpion, fish—wrapped around a sphere. The year that I lived in Florence with my family, we'd go to the park on Sunday afternoons in the fall and hope that the carousel was open. The crispness in the air and the setting November sun were melancholy, and to catch a glimpse of the wheeling carousel, with its string of bright white lights and tinsel music, was like swallowing something warm: a tonic against the encroaching frost, a draught of heat to counterbalance the chill in our cheeks and fingers. My four-year-old would hoist herself up on some slick, colorful beast, and I'd take a seat in one of the swan-boat style benches painted with curlicues and flowers, my one-year-old perched on my lap, and round we'd go.

Today, as cancer cycles through my body, the memory feels like a talisman, part charm, part warning: tiny spinning animals, earthbound specks of light beneath the darkling sky.

The hospital where I get my chemo infusions is brand new, bright and white and cheerful. This is encouraging. If my body is bent under the weight of its sickness, the building at least is upright, seemingly immune to the battering of the elements and the wear and tear of time. I grab my café au lait from the kiosk in front of the hospital. The automatic glass doors part before me and, from behind a modishly curved white desk at the entrance, a young woman smiles expansively and wishes me welcome. I feel as though I could be at a bank or a law office, depositing a check or filing my tax returns, which, I suppose, is the point. The modern Western biomedical project is designed to put nature, red in tooth and claw, to rout. That most obscene and inhumane of natural outcomes, death, is an enemy not to be tolerated. We see this in the military vocabulary commonly used to describe cancer treatment itself: it is a "battle," always "valiantly fought," and the patient who "beats" cancer in this zero-sum game is reckoned a "survivor": humans 1, nature 0.

The blood draw station on the ground floor is my first stop. I wait for the nurse to call my number—0756—then take a seat in the phlebotomist's chair. First the squeak of the rubber strap as she knots it around my bicep, then, "Make a fist." I wait for what I know will come next: "Good veins," she murmurs from behind her surgical mask, and I feel a blush of idiot pride, as I always do: gold star for anatomy. My siphoned blood is whisked offstage for testing, and a specialist trained in the modern art of blood-reading will decipher my fate. "Your numbers look good, Ellen," my oncologist will later assure me.

At the blood draw station, they play ambient environmental music on a loop, the kind that mixes synthesizer pan pipes with recorded nature soundscapes: trickling water or wind or bird calls, sometimes all three. It is music I've only ever heard in yoga classes or in the massage rooms at boutique spas; it makes me think of warm towels and the sweet smoke of incense. I learn that MRI imaging studies show the brain relaxes its fight-or-flight response when exposed to nature sounds, even artificial ones, and presume this is the logic behind the piped-in soundscapes. Still, despite these therapeutic findings, I find the music incongruous as I look out at the other patients in the waiting room. We are so many scattered atoms, shuttled one by one through this clean, white, biomedical machine. What have we to do with wind and water and birdsong, the bosomy embrace of nature?

After I'm cleared by my oncologist, I head up to the fourth floor and check in. The waiting room crowd here is a little more haggard than that down below: an older man in faded jeans coughs from behind his white protective face mask; a woman in a flowered headscarf pages through *People* magazine. This level is for the Truly Ill, the transplant and cancer patients. My nurse Michael comes out to get me. He is tousled-blond and chipper, tall and broad shouldered, with gym-perfected pecs and biceps that swell to fill his red scrubs like a hand fitted snugly in a glove. We chat about small things as he preps my medicines. Did I try the Tylenol and Claritin mix the doctor recommended for post-chemo bone pain? Yes, I say, and it doesn't do a damn thing; only Advil helps a little. Michael nods vigorously in agreement. "I'm an Advil fan too," he confesses. As he slides the IV needle into my arm, he apologizes with a wince. "Liiiittle pinch, I'm so sorry."

In the infusion room, each patient gets her own minicabin, discreetly partitioned off from one another; we sit side by side, as if in a movie theater, facing floor-to-ceiling windows that run the length of the building. We each have our own private room with a view. This is a university hospital, and across the street they are building a new dormitory for medical students.

Beyond the construction is a train track, then the 101 freeway, and beyond that the dusty green foothills of the San Gabriel Mountains. Hawks circle above the palm trees, searching out snakes and mice and rabbits and baby birds, as humans scuttle about on the sidewalks below, noses buried in their phones.

We no longer see the earth as part of us, an entanglement of heart and sinew, bound by the shared tempo of circling blood and circling planets. Rather we exist side by side with nature, an uneasy cohabitation, each carving out its own separate sphere. Nature is ingenious and tries to adapt to our incursions. The hawk learns to hunt and pick across an urban landscape: it gorges on human-fed songbirds, a fattened luxury unknown in the wild. It learns to nest in eucalyptus trees, good in a pinch when more hospitable native trees are scarce. As for humans, we take nature in medicinal doses, and always on our own terms, packaged into easily digested health supplements like the synthetic sounds of birds and water.

In the vegetable kingdom, there are few plants more toxic to humans than *Taxus baccata*, or the European yew tree. Ingesting even a handful of its bright red berries or evergreen needles can be lethal. They contain compounds known as taxines that inhibit calcium and sodium transport in myocardial cells, leading to arrhythmia and then heart failure.

The yew is one of two plant ingredients that the witches in Shakespeare's *Macbeth* throw into their maleficent brew: to accompany "root of hemlock digg'd i' the dark," they add "slips of yew / Silver'd in the moon's eclipse." The animal ingredients the witches harvest for their potion sound even more monstrous to the modern ear: "Fillet of a fenny snake, / In the cauldron boil and bake ; / Eye of newt and toe of frog, / Wool of bat and tongue of dog." When, upon returning from battle, Macbeth chances upon these hags in the forest and asks them to riddle his future, we are not surprised that his luck turns bad.

But early modern readers would have recognized many of these baleful-sounding ingredients as standard stock in medicinal recipes, if not actually available at the apothecary's shop. Herbs and minerals and animal parts were key to concocting all manner of health-inducing pills, oils, and ointments. Edward Topsell's 1658 *History of Four-Footed Beasts, Serpents, and Insects* recommends, "among the medicines arising out of the female goat," "that the right eye of a green living Lizard, being taken out, and his head forthwith struck off, and put in a Goats skin is of a great force against . . . Agues," and that a freshly killed adder, "included in a pot with the scrapings of Vines, and therein burnt to ashes," makes an effective paste for soothing boils.

So thin was the line separating the medical from the black arts, that Robert Turner's 1664 herbal manual begins by praising God for enduing "the Plants and

Grass of the Field with such salubrious Faculties for our health and preservation," but then quickly turns to warn against their improper use. He speaks ominously of "druids," "heathens," "Medean hags and sorcerers" who forego the lawful use of God's creatures and, instead, "out of some Diabolical intention, search after the more Magical and occult Vertues of Herbs and Plants to accomplish some wicked end."

There was sinister, demon-ish magic at work in the witches' brew. But the protoscientific healing arts, freely mixing alchemy, astronomy, and botany, were magic too: christened "natural magic" from the Latin *magus*, one practiced in the art of controlling or predicting nature. Healers worked by trial and error to find the exact combination of herbs and stones and animal parts that, properly compounded, would seal the rent in the universe that was a sickly, unbalanced body. The empirical sciences of the seventeenth century arose, not in opposition to natural magic, but as its outgrowth; they had their source, not only in the arid air-castles dreamed up by university philosophers, but also in the earth-caked grit of the apothecary's lab.

Biologists in the 1970s isolated a molecule in the bark and needles of the Pacific yew—paclitaxel, related to but distinct from the lethal compound taxine B—that prevents cell division in tumors. In order to divide, cells need to be supple and soft, able to split like a large hunk of bread dough twisted into two smaller knots. Paclitaxel stiffens the cells' molecular spines so that when they go to divide, they splinter apart instead, scattered into lifeless pieces like Hera's crushed crab. Patented and lab-produced as Taxol, a version of this tree serum is one of the chemical concoctions that I receive, intravenously, every three weeks. I know first-hand about the Janus-faced yew tree: balm or bane depending on its use.

I have always been bad at numbers. In eleventh grade Chemistry, I sat next to another mathematical dunce, a shy, sandy-haired girl named Tina, and the teacher used to pair us up to solve the equations he'd chalked out on the board. "Maybe between the two of you, you'll come up with half the answer," he'd quip. Today, with my unpracticed chemist's eye, I scrutinize the formulae for poison ($C_{33}H_{45}NO_8$) and salve ($C_{47}H_{51}NO_{14}$); I trace the clusters of their rings and sticks with my finger. They look like miniature constellations, printed in black and white. One arrangement of atoms stops the heart. But scramble the order, add or subtract an element, and it stalls a tumor. This is a species of combinatory sorcery, mystery at the heart of matter.

After Michael hooks me up, I sit nodding beneath my chemo IV tree, a sac of Taxotere hanging from its metal branches like a swollen plastic fruit. I imagine the matter-rich, earthy origins of this elixir—the roughness of scraped brown tree bark and crushed needles, boiled and baked into a clear quintessence—and suddenly, the self-enclosed loop of my body splits open like a vein and floods into the larger wheel of natural life, mixing my blood with its blood, a universal tumble and whirl, the stars in my cells. Slips of yew circling through my tissues, carrying out their chemical magic in the blind heart of my flesh, binding and unbinding, ever-shifting atoms blooming into momentary constellations that (with a dash of luck) will right my body's wobbly universe, at least for a time.

THE ATAVIST

Right after my sister gave birth to her first child, the doctor smiled and congratulated her, and then said in a tone suddenly turned more serious: "Now don't be upset, but there *is* a small tail." My sister was speechless, torn between panic that her firstborn was an evolutionary throwback and annoyance that the doctor thought he could soften the blow by qualifying the tail as "small." She had heard incorrectly, of course; what the doctor had actually said was "there's a small *tear*," a perineal rip requiring a few stitches before she could hold her newly swaddled infant (mercifully *sans* tail) to her chest.

Yet who among us has ever given birth without suspecting, at some point along the way, that she was sliding back into a primeval, animal mode of being, at once beneath and beyond the human? The birthing body—all pushing, huffing, howling, seeping, spewing, splitting—is a violation of the self and its borders so total that what is remarkable is not the relapse into dumb creatureliness, but rather just how quickly we pull ourselves back from the animal brink once the act is done. How instinctually we reprise our human selves, redon the cloak of seemly separateness, of decorous containment.

My sister's baby didn't have a tail. But in that twilight world of birth, of churning blood and mucus and sweat and shit, turning inside out, soft parts squeezing, skin birthing bone, mine, its, ours—within this mad thud and swirl of flesh, I ask: would such a relic of our animal origins really be so out of place?

The Victorians didn't think so. In his 1844 *Vestiges of the Natural History of Creation*, amateur geologist Robert Chambers put forth the speculative hypothesis that all creatures, from the lowly protozoan pulsing in the sea to the noblest human form, were evolutionary links in a divine chain of creation. "[God] has chosen to employ inferior organisms as a generative medium for the production of higher ones, even including ourselves," he speculated, and all creatures have "assigned to them by their Great Father a part in the drama of the organic world." Fifteen years later, Charles Darwin's *The Origin of Species* made animal-human kinship a household topic of conversation. "The great Tree of Life," he observed, "fills with its dead and broken branches the crust of the earth and covers the surface with its ever branching and beautiful ramifications." One could read the geological layers of the earth like a book, from mollusk to mammal to man.

Yet neither Chambers nor Darwin believed evolution was unidirectional. Didn't the very existence of a pathway leading from animals to humans suggest devolution was a possibility as well? Darwin's broken, buried branches were not so much dead as dormant, lurking in the roots of our being and threatening chthonic resurgence when we least expected it. Chambers cites those well-known cases where human babies are born with underdeveloped organs. "The heart, for instance, goes no farther than the three-chambered form, so that it is the heart of a reptile," he explains; some fetus hearts even stall in the "two-chambered or fish-form." "Thus we see nature alike willing to go back and to go forward," he cautions, and a poor environment, within the womb or out of it, leads to "an unequivocal retrogression towards the type of the lower animals."

The specter of regression haunted the fin de siècle, from science fiction to sociology to psychology. Freud himself proposed that all the civilized values humans hold dear—honor, duty, faith, love of family and country—were the agonizing, ever-fragile work of keeping the seething id and its animal instincts in check. These repressed forces stood always at the ready, waiting for their chance to come surging back.

H. G. Wells's 1896 novel, *The Island of Dr. Moreau*, is a typical period fantasy exploring the beast within. The narrator Edward Prendick, sole survivor of a shipwreck off the coast of Peru, is picked up by a schooner and nursed back to life by a doctor who happens to be onboard, a man named Montgomery. If there is something oddly off-putting about Montgomery, with his "watery grey eyes" and "dropping nether lip," stranger still is the doctor's servant: a "misshapen man, short, broad, and clumsy, with a crooked back, a hairy neck, and a head sunk between his shoulders." When he turns to face the narrator, the flash of his face is shocking, "forming something dimly suggestive of a muzzle, and the huge half-open mouth showed as big white teeth as I had ever seen in a human mouth."

When Prendick is strong enough to take a trip up on deck, he is confronted by what looks like an "ocean menagerie": a "number of grisly staghounds" fastened by chains to the main mast; "a huge puma . . . cramped in a little iron cage"; a number of hutches containing rabbits, and finally "a solitary llama . . . squeezed into a mere box of a cage." When the drunken captain of the ship comes bellowing and staggering up from below deck, he appears only marginally more civilized than the pack of mangy beasts or Montgomery's strange companion. This is a world, Wells intimates, where the line between man and beast has come close to collapse.

The narrator and Montgomery dock on a remote island where the eponymous Dr. Moreau, hounded out of London for the cruelty of his animal experiments, has retreated to continue his research in secret. Moreau's goal is to build a rational creature from scratch, grafting and stitching together live animal tissues—leopard, ape, llama, swine—in a bid to reroute evolution. The mugging man-ape aboard the schooner was just one of hundreds of such hybrid creatures to emerge from under Dr. Moreau's knife in his quest "to find out the extreme limit of plasticity in a living shape."

When the Beast People go berserk and kill both Montgomery and Moreau, Prendick manages to escape with his life. But he remains haunted for the rest of his days by Moreau's botched experiment in artificial evolution. Once repatriated, he cannot shake the suspicion that the humans he encounters on the bustling streets of London are really just Beast People in disguise, "animals half wrought into the outward image of human souls, . . . [who] would presently begin to revert,—to show first this bestial mark and then that."

In seeking to explore the terrors of evolutionary backsliding, Wells needn't have searched so far afield. Monstrous regressions could be found much closer to home, albeit in sadder, more mundane form, in the bodies of cancer patients. In 1889, three years following the publication of *The Island of Dr. Moreau*, surgeon Herbert Snow delivered a lecture entitled "The General Theory of Cancer-Formation" at a London hospital. In it, he explains how dividing cancer cells take on a "quasi-independent vitality" within the sufferer's body, "refusing allegiance to the laws which regulate the rest of the system." "These tiny creatures," he notes, "revert to their primary amebiform condition, and live, not only on a separate footing from the remainder of the animal, but even at the cost of the latter."

I saw Snow's "tiny creature" once, projected up on an ultrasound screen. The doctor waved a gel-slicked wand over my breast, and the interloping mini beast came in and out of focus: barely decipherable—a scrawl, a scratch, a malignant insect curled and crouched inside my flesh.

Nearly a century and a half after Dr. Snow launched his regression hypothesis, some scientists claim he may have been on to something. The behavioral similarities between cancer cells and elemental forms of life—single-celled protozoa, minimally organized cell colonies—suggest that cancer may indeed be understood as a species of atavism, the return of "ancestral cellular functions regulated by genes that have been largely suppressed for more than 600 million years," according to physicists Paul Davies and Charles Lineweaver. Early life forms had only one strategy to evade environmental threats to their survival: "to up the mutation rate until a solution emerged." When a cell senses, rightly or wrongly, that something is "off" in its environment, it may default to the archaic, hardwired survival strategy of rapid mutation. A malignant tumor, viewed in this evolutionary perspective, is a species of living fossil, and cancer not just a disease, but a "competitive struggle between a battle-scarred protozoan from the Pre-Cambrian" and its metazoan host.

It stands to reason that cancer treatments, from surgery to chemotherapy to radiation, are so many efforts to recreate in miniature the ecosystem stressors that threatened the survival of early life-forms in the past. Radiation mirrors the atmospheric conditions of the depleted ozone layer on Precambrian earth. The estrogen blockers I swallow daily are meant to catalyze a food chain collapse. My body is a tiny earth, and my doctors are trying to engineer a targeted mass extinction of a life-form teeming in and upon it. The trick is to drive the niche population that is my tumor out of existence without blowing up the whole planet.

So, I live with this freak, this fossil in my flesh, flashback binding me to time before time: swarm of cells going bump in the Hadean dark.

Therein lies the paradox: all of God's green earth exists only because this early protozoan was a mercenary son of a bitch. It took no prisoners; doubled, quintupled, centupled in a mad dash to keep one step ahead of the ever-fickle environment of the ancient earth. The quasi-immortality of this lump of cytoplasm is the rock upon which Darwin's fabled Tree of Life took root and burst into flower, burst into birds and bees, burst into me.

So now this microscopic ancestor, antediluvian grantor of my being and the being of all things that creepeth upon the earth, is back. And really, what can I say?

To be dethroned by this ungainly ancestor-progeny, inventing its way to a future that doesn't include me, is instructive, in its way. It puts me beside myself, forces me to hold myself at naught. Zero, *nihil*, no-thing: this sharer of my flesh casts me aside but also puts me in my place. Cancer reveals an order of created being that does not hold me at its center. "It's nothing personal," my tumor seems to say and, well—that's for sure. Cancer's utter disregard for the survival of your person dips you in the fire of the impersonal, which is at once its horror and its peculiar enlightenment. It reveals your participation in a cosmos that vastly exceeds you and your paltry blip of a species, let alone your individual life.

King Lear knew what it was like to be ousted from the castle keep of human personhood. Tossed out of doors by his thankless daughters, he reproaches them with all manner of animal epithets: they are vultures and kites, pelicans and wolves, sharp-toothed serpents—a gnashing monster menagerie of claws and fangs and beaks.

Yet it is only once he descends to animal being himself, once he has "abjure[d] all roofs," become "a comrade with the wolf and owl," that Lear discovers what it means to be human. Reduced to unhoused flesh, Lear is able for the first time to feel compassion for the suffering flesh of others—and this is his salvation.

To make peace, then, with the poor, bare, forked animals that we are is not to take a step backward. On the contrary: it is to uncover something sacred, to spy the ground of being, the hidden spring and pulse of life bubbling at earth's green heart.

QUARTZ CONTENTMENT

Scrolling through Twitter one day in the very early weeks of the COVID-19 pandemic, I came across a tweet that struck me as equal parts mirthless and true (the two often go hand in hand). "Things will be fine, eventually, in thousands of years, for rocks," quipped comedian Donni Saphire, garnering 91.4 thousand likes. It reminded me of a saying my mother used to habitually trot out when I was growing up, whenever I got exercised by some trivial contretemps or other—bad hair day, missed party, hallway snub. "In the grand scheme of things," she would drawl, as if to underscore the grandness of the time scale envisioned, "it just doesn't signify." This was patently untrue, of course, and irritating to boot. For the teenager, as for the toddler, there *is* no grand scheme of things; there is only the absolute *now*, and it signifies absolutely.

Still: everything *will* be fine, in the grand scheme of things, for rocks. Why not, then, in this era where we find ourselves locked in a perpetual calamitous stutter, teetering on the edge of catastrophe—why not try to imagine things from the unmoving, diamond-hard perspective of the mineral kingdom? It couldn't hurt.

I'm not the first to suggest it. Emily Dickinson, who had an exquisite sense for the grand scheme of things—the infinitely small to the infinitely large—frequently turned to mineral imagery in her meditations on time and mortality. Stratigraphic studies in the emerging science of geology in the nineteenth century revealed previously unimagined time scales that took the breath away. Natural scientist and Amherst College chemistry professor Edward Hitchcock was a close Dickinson family friend, and his 1840 treatise *Elementary Geology* was assigned reading at the school Emily attended. One contemporary geological primer writes of the "great convulsions" that shaped the ancient surface of the planet, noting that these early epochs "may seem of almost inconceivable duration" when compared with "the ephemeral existence of man on the earth." Victorians were left staring into the archaic abyss of the earth's origins, pondering their smallness. No accident, then, that Dickinson's poetry should be studded with mineral outcroppings—quartz, granite, alabaster, adamant—all witness to her geological interests.

Of course, poets have always used stones to convey the insensate, mute quality of the dead. But in speaking of death, Dickinson resorts to stone imagery more

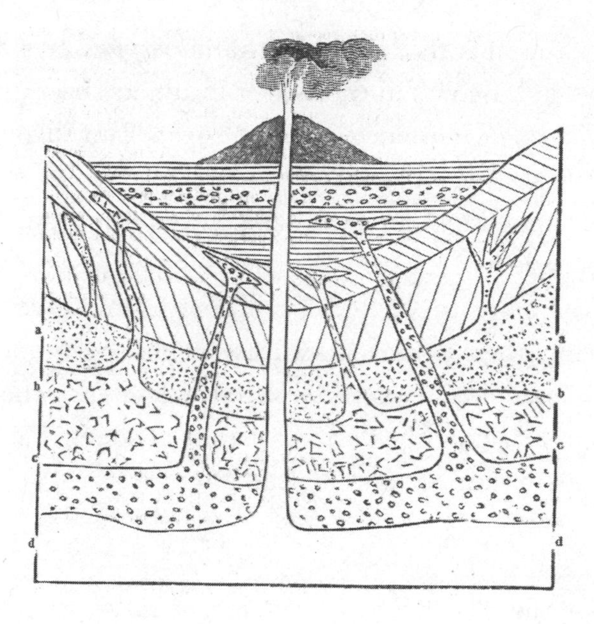

consistently, more creepily, and more *literally* than perhaps any other poet in the English language. Poem 519, "'Twas Warm—at first—like Us," for example, is a forensic description of a body in the process of rigor mortis, transmuting from person to thing: first the "Forehead copie[s] stone," then the eyes congeal like a "Skater's Brook," until the body "drop[s] like Adamant" into the grave. The corpse's "multiplied indifference" is given a more cheerful spin in "Safe in their Alabaster Chambers," where Dickinson imagines the dead as so many "untouched" sleepers tucked safely in their stony beds, insensible to the dizzying arc of interstellar time and the rise and fall of human civilizations all around them: "Diadems drop and Doges surrender, / Soundless as dots on a disk of snow."

Dickinson is fascinated by the imperviousness of stone—its passivity in the face of insult, its sheer uncomplaining persistence across the ages. "How happy is the little stone / That rambles in the road alone," she writes. Inured to "care" and "exigencies," the stone is "independent as the sun." It is hard to imagine she did not have Hitchcock's geology primer in mind when she wrote these poems. Of what possible significance is the span of a human life, she seems to ask, when measured against the vast swathes of uncounted and uncountable time at granite scale?

Among the side effects of that class of antidepressant drugs known as SSRIs—selective serotonin reuptake inhibitors—is what psychologists call "blunted" or "flattened affect": a diminishment in the range of emotive expression available to the patient. I have been on and off—but mostly on—SSRI medication since my grad school–induced nervous breakdown and first depression diagnosis at the age of twenty-four. In other words, for upward of thirty years. When I first started taking the medication, it had the effect not of suppressing my emotivity altogether, but of simply overriding its paralyzing force. No longer was I so panicked that I couldn't leave the couch; no longer did I cry so much I couldn't get out of bed. But then, as the years wore on, I noticed that I was, in fact, less apt to feel than I had been. Tears that had once rolled so abundantly came infrequently, if at all. Where once worry as to the fate of my soul (as a child) or my sanity (in my teens and twenties) had consumed me, with time I grew increasingly unable to feel any kind of way at all about the future, at least when it came to my own individual person. When I looked toward the future, I did so without any marked feeling of desire or apprehension as to what it might bring— not unlike Dickinson's stony sleepers, "Untouched by Morning/ untouched by noon."

To be fair: even before the Prozac I had not been a person given to passionate intensity. Excessive emotion was frowned upon by both sides of my family, though for different reasons. My maternal grandmother met every one of life's obstacles with a stoic cheerfulness that was at times almost eerie. While never expressly forbidding my mother and her sisters from having feelings, she nonetheless coached them to smile politely through pain, whether physical or emotional. After my grandmother had a bad fall in her eighties, an X-ray revealed, to everyone's great surprise, that she had once broken her back—an event she either never registered at all or, having registered it, thought too impolite to mention.

My father's parents, on the other hand, exhibited an apparent allergy to affect of any sort, including and perhaps especially when it came to expressing emotion for their children. Following a popular 1930s parenting manual touting the salutary effects of exposing infants to "bracing" cold weather, my grandmother would place my father in his pram and leave him out on the front porch for timed intervals during the winter, an icy practice that, while in no way diminishing the love I know she felt for him, nonetheless hints at the difficulty she often felt in expressing it. As for my grandfather, a lifelong bird-watcher, I was flipping through his notebooks one summer and fastened on a passage where he described a particularly impressive hawk sighting off the Delaware Water Gap. He devoted half a page to its breathless description. Only later did I realize the occasion for his having been in the Delaware Water Gap at all: he had just deposited my fourteen-year-old father off at prep school. But of this parting from his eldest son, not a word; my father reduced to an extra-textual ghost.

Still, above and beyond whatever genetic tendency toward affectlessness I might have come by naturally, I believe the Prozac did have an additional numbing effect. The air of neutral indifference with which I appeared to approach my own life became a topic of some medical curiosity when I was diagnosed with breast cancer in the fall of 2019. Among the many doctors I consulted in the wake of my diagnosis was a psychiatrist appointed to check in on how I was coping mentally with the prospect of mastectomy and chemotherapy. I have a clear memory of the consult with this officious, brisk young doctor with chunky clogs and a clipboard. I rattled off all the usual psychiatric history highlights while she nodded and scribbled notes. "But how do you *feel*?" she pressed. I assured her that I was coping fine with the diagnosis. "I feel fine, really," I kept repeating, smiling apologetically, aware that something in my affective response to the falling-apart of my own body was falling short of what she expected. When I read my clinical report afterwards, I found this: "Patient seems to be speaking with some isolation of affect that is noticeable (discussing her diagnosis and sensitive topics with little to no emotional reactivity)" and, a bit further down in the report, under the heading "Affect": "Largely congruent but a bit blunted, limited range."

I found her assessment fair enough; I knew that the general lack of curiosity and worry with which I greeted my diagnosis was probably atypical. But one still might pose the question: lacking reactivity—blunted, limited range—*by what metric?* Compared to whom, or what? Measured by what zenith of acuity, sensitivity, or self-concern had I been found wanting?

A capacity for sensation, or what my doctor called "reactivity," is among the oldest and most-trusted philosophical criteria by which to judge a creature's place in the hierarchy of living things. Aristotle famously created a taxonomy of "souls" (*psyche* in Greek; *anima* in Latin) to describe an ascending biological scale from vegetable to animal to human. Vegetables were capable of growth and reproduction, which Aristotle called a "nutritive" soul. Animals, one notch up the ladder, exhibited all the ensouled properties of plants but in addition were capable of feeling, motion, and digestion. Finally, humans topped the chart as the only living beings possessed of a capacity for thought, or what Aristotle called a "rational soul." Minerals—inert, capable of neither sense nor motion—fell outside the scope of life altogether.

The taxonomy stuck, more or less, and when the mania for scientific categorizing really got going in the eighteenth-century, Carl Linnaeus would formalize Aristotle's theories by dividing the world up into animal, plant, and mineral kingdoms. He even coined an easy-to-memorize formula for aspiring naturalists: "Stones grow; plants grow and live; animals grow, live, and feel." Reading the psychiatrist's report, I saw myself slipping down the rungs of the Great Chain of Being: past the animal, past the vegetable, landing with an adamantine thud among the minerals.

Yet what if, like Emily Dickinson, we could teach ourselves to entertain the possibility of a nonhuman scale—a geologic scale—as another way of looking at the world? Writing to her brother Austin in the early spring of 1848, eighteen-year-old Dickinson alludes to the chemistry class she was taking at neighboring Mount Holyoke Female Seminary: "Your letter," she quips, "found me all engrossed in the history of Sulphuric Acid." From a young age, she demonstrated a penchant for viewing history from the perspective of the inorganic: rock, atom, acid. Dickinson is often qualified as a "morbid" poet, and in many ways she is. But if she is obsessed by the instant when unfeeling death seals human sight—"And then I could not see to see," she writes in one of several poems that seek to capture consciousness on the brink of disappearance—it is perhaps because, in this closing down of human perception, she senses the opening up of an altogether *other* way of being in contact with the world.

Dickinson ascribes to the dead the geological insensibility of granite, yes. But she also employs the point of view of rocks to approximate certain interior mental states she experienced *while still living*: periods she felt to be a kind of death-in-life, as though through some diabolical trick, God had slipped a hard crystal into the place where once a human consciousness hummed. In "After Great Pain," Dickinson's narrator describes a suspended state of frozen torpor that seizes her in the aftermath of grief. The narrator moves through life "mechanically," having lost all sense and feeling for her surroundings, "Regardless grown, / A Quartz contentment, like a stone." In "It Was Not Death," Dickinson narrates from the impossible point of view of what she calls chaos itself: a field devoid of the usual markers by which humans orient themselves in space and time, stripped of the minimal conditions necessary to establish a perceptive, feeling, thinking relationship to the world at all: "But most, like Chaos - Stopless - cool - / Without a Chance, or spar - / Or even a Report of Land - / To justify - Despair." The poem conjures an inert, watery waste, "without form, and void," before God's breath and hand had created the light, shape, and dimension through which we recognize our familiar human-centered world. Such impersonal mental states—quartz contentment, chaos cool—were clearly terrifying for Dickinson. But they were also terribly instructive, and she seized upon them as apertures through which we might catch a glimpse of the world without us.

The stony perspective Dickinson brought to her poetry can be found in other fields as well. Early twentieth-century Russian mineralogist Vladimir I. Vernadsky first coined the term "biosphere" to refer to that thin envelope of life encircling the earth. Studying life through a rigorously geochemical lens, his writing is bracing in its refusal to grant humans a privileged place in the cosmic scheme of things. He flips the inherited scenario that a rocky, cooling earth served as mere backdrop to the main event of life's emergence. Instead, Vernadsky refers to the entirety of "living matter"—including its human variety—as essentially one giant, interlocking "energy transformer" whose metabolic processing of chemical and energy inputs accomplishes the "colossal geochemical labor" of distributing matter and solar energy across the planet. We might use minerals—mine them, transform them, bend them to our will—but in the grand scheme of things, or glimpsed through a thermodynamic lens, the elements in minerals are also *using us* as the universe pushes inexporably toward entropy. The difference is simply one of perspective.

"Life begets rock, rocks beget life," comments minerologist Robert Hazen. Minerals and living organisms are coevolving, fully intertwined, with the majority of today's five thousand–plus documented mineral species a result, in one way or another, of the 3.8 billion years of biological activity on the planet. Some of the most baroquely beautiful crystals in existence form through the oxidation of copper sulfide minerals; these crystals only became a chemical possibility once the evolution of algal photosynthesis had flooded the earth's atmosphere with oxygen two billion years ago. Malachite, for instance, with its grape-cluster swirls of heart-stopping green, is the indirect result of this primeval biotic effusion. On the organic side of the equation, early invertebrates harnessed aragonite and calcite crystals from the ocean, folding these minerals into their own metabolic cycles to build teeth, bone, and shell. No matter how one looks at it, animal and mineral destinies are linked.

When I told a friend about my incapacity for future-thinking or worry, my peculiarly dissociated approach to my own life, he said, "Isn't that just another name for wisdom?" "Wisdom literature" is, indeed, often touted as wise because it urges readers to ponder questions of scale, the transitory nature of any single life in the grand scheme of things. "One generation passeth away, and another generation cometh: but the earth abideth for ever," we read in Ecclesiastes.

Wisdom or chemical lobotomy—sagacity or brain deficit—who is to say? In the meantime, I am interested in what I might *make* of the peculiarly quartz-like lens through which events present themselves to me. To see like a stone, in Emily Dickinson's sense, is not to turn a cold shoulder to the suffering of a sentient earth and the beings that populate it. On the contrary: it is to fine-tune our attention to metabolic entanglements happening far beyond our knowledge. It is to sense those grand scale scoops and arcs that bind together the atoms of the cosmos, including—but no longer reduced to—our own species' small, borrowed parcel of stardust.

LAPIDARY MEDICINE

It was winter in Los Angeles, I was feeling glum, and rose quartz was having a moment. In February, the spa at the Ritz-Carlton hosted a month-long Valentine-themed event called Everything Is Coming Up Rose Quartz; interested wellness-seekers could soak in the physical and spiritual healing benefits of the pink crystal in the form of a facial, full-body massage, or mani-pedi. The spa was partnering with the skincare company Knesko, whose line of Gemclinical facial products, according to their website, employ select precious gemstones and minerals—Gold, Pearl, Diamond, Amethyst, and Rose Quartz—to "activate your chakras and rejuvenate skin." The gold-infused products, for instance, "align with your sixth chakra, or third eye chakra, to balance your intuition," while Rose Quartz "aligns your fourth chakra, or heart chakra, making it easy to give and receive love."

I was intrigued. I mean, who *doesn't* want to imagine themselves as a crystal, all liquid light and angled beauty, impervious to the ravaging wear and tear of life, and with chakras aligned into the bargain? And then, ever since moving from the East Coast to Los Angeles in 2012, I had been fascinated by the pervasive, affable New Age-y quality that I encountered in even casual daily interactions with the city's natives. At PTA coffees, I eavesdropped on conversations that slipped effortlessly from auras and horoscopes to math scores and Secret Santa gifts. This was my chance, I told myself, to do an anthropological deep dive into spiritual LA. I couldn't afford the Ritz-Carlton, but I found a cut-rate option in East Los Angeles where the crystal experience was on offer for a fraction of the Beverly Hills price. I booked an appointment.

Arriving at the spa, an unassuming, low-slung bungalow in Silver Lake, I walked through a leafy courtyard and into a little warren of wooden rooms clustered around a burbling stone fountain. The walls were hung with sand- and chocolate-hued swathes of fabric, the open space accented with wicker furniture and yellow hemp ottomans. The esthetician—whose skin was, as the receptionist had assured me, so glowing that it appeared to light up the dim cubicle in which I found myself—then commenced the facial. What followed was a bewildering, fast-paced suite of orange blossom cleansing creams and verbena oil and whipped deep-moisture masks and peptide serums and a half-dozen other preparations that I eventually lost track of. A steam-emitting nozzle coated my face in a slick layer of hot droplets as the esthetician massaged first one product then the next into my skin with the tips of her fingers. At one point she placed a plastic dome over my head for a bask in an "anti-aging" LED light, and the backs of my eyelids glowed orange even through the protective eye patches.

The crystal portion of the whole production came at the end and consisted of a rose quartz roller facial massage and a spritz of atomized rose quartz–infused water. The esthetician effortlessly navigated the twin rollers—like miniature paint rollers but made of stone—across my forehead and down my cheeks, up and over the bridge of my nose, and all around my chin, in a series of brisk, silky circles. The stone was cool but not chilled, and the rollers occasionally made a pleasing clicking noise when they chanced to meet up in the middle of my face, like a pair of die shaken in a cup. I asked the esthetician if she refrigerated the rollers, and she shook her head firmly in the negative. "Rose quartz—I mean *real* rose quartz," she clarified, "is naturally cold." As a final touch, she sprayed my face with "rose quartz and mineral enriched toner," which hung in the air for a split second before falling onto my skin like a peppering of tiny, soft fireworks.

I don't know if I felt more balanced or rejuvenated after my facial, and my chakras felt pretty much unchanged. It was perhaps to be expected; the spa's skincare products were, after all, but a down-market version of the Gemclinical® brand. They made no grand claims to being charged with Reiki energy. But I had to admit that the rose quartz rollers were cool.

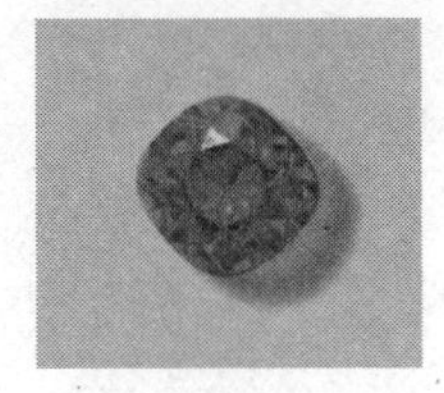

Belief in the physical and spiritual healing benefits of stones has a long history in the West, from the earliest Greek philosophers through the encyclopedic natural histories and medical treatises of the Renaissance. The first-century Roman polymath Pliny the Elder devoted the final book of his encyclopedic *Natural History* to precious stones, describing their origins, appearance, and reputed medicinal properties. The diamond, or what the ancients called *adamas*, "overcomes and neutralizes poisons," he reports, "dispels delirium, and banishes groundless perturbations of the mind." An agate stone, held in the mouth, allays thirst. Amber, ground into powder and mixed with rose oil and honey, cures earache. There is "no green in existence of a more intense colour" than emerald, Pliny asserts, and its "soft green tints" offer a tonic for strained or tired eyes.

The medieval Christian writers who inherited these early stone encyclopedias—called *lapidaries* from the Latin *lapis*, or stone—were leery of their pagan magic, but skirted the problem by cloaking classical stone lore in a layer of biblical symbolism. Christian lapidarists poured the old wine of pagan stone wisdom into the new skins of scripture, reorganizing their material to fit the twelve stones that decorated Aaron's breastplate in Exodus or the gems paving the heavenly city of the New Jerusalem in Saint John of Patmos's apocalyptic vision. According to one thirteenth-century French lapidary, sapphire, as the second of the twelve stones lighting up the new Jerusalem, corresponded to the second heavenly virtue, hope; contemplation of the gem's sky-blue depths would accordingly "raise men's souls to the contemplation of the heavenly kingdom." By virtue of their placement in a sacred web of divine correspondences uniting heavenly and earthly realms, stones functioned as material gateways to spiritual salvation.

By the fifteenth and sixteenth centuries, stone wisdom had evolved into the pseudoempirical science of lapidary medicine. Medical practitioners traced the elemental "sympathies" that bound human bodies to stars, plants, animals, and stones and applied the homeopathic principle of "like cures like" to heal their patients, offering up medicinal recipes to remedy everything from toothache to delayed labor. One popular 1550 medical handbook proposed that a paste concocted of powdered red coral would hasten the removal of a rotten tooth, as though through the sheer force of sympathetic resonance—a red gum matched to a stone of the same hue—a body could not help but inch itself toward cure. By the same logic an eagle stone—a type of geode containing a stone within a stone—could be held against a laboring woman's belly to induce a baby to descend the birth canal, like a magnet pulling iron. To heal, in the lapidary tradition, was to believe that poetry could transform flesh.

Today's Los Angeles crystal culture bears only faint traces of the Western lapidary tradition. It is, instead, much more recognizably New World in its breezy promiscuity, gathering up tidbits of spiritual wisdom from every corner of the globe (and even from outer space) and lumping them all together. The early origins of today's SoCal crystal cult might plausibly be traced back to the early 1900s when the American branch of the Theosophical Society migrated west. Theosophy was the brainchild of Russian-born clairvoyant Helena P. Blavatsky who, in 1877, published a rambling and vaguely conspiratorial thousand-page manifesto entitled *Isis Unveiled: A Master-Key to the Mysteries of Ancient and Modern Science and Theology.* In it, she claimed that in her extensive travels East, visiting "deserted sanctuaries" and hobnobbing with "the sages of the Orient," she had uncovered an ancient wisdom cult long suppressed by Western materialist science, whose secrets she was now about to unfold to willing initiates. Early twentieth-century American adepts of the movement found a welcome home in the then-frontier town of Los Angeles, where in 1912 they founded the colony of Krotona on ten acres of land in Beachwood Canyon, in what is today the heart of old Hollywood. Fueled by wealthy Theosophist patrons, before long the canyon had blossomed into an *Arabian Nights* mishmash of lotus pools, onion domes, and Moorish arches, planted with suitably "exotic" citrus and olive and pepper and palm trees, like a white settler fever dream vision of the Orient—which, of course, is precisely what it was.

From the beginning at Krotona—and in this, the Theosophists preceded the Beach Boys by at least a half-century—picking up good vibrations was the heart of the matter. "There is one word, vibration, that is becoming more and more the keynote of Western science," noted Theosophical Society president Annie Besant in 1901, "as it has long been that of the science of the East." Just as the eye and ear are primed to pick up light and sound vibrations, so the pineal gland—an organ still in the process of evolution—can be trained to pick up ambient thought vibrations as they travel through the ether, Besant asserts. If one focuses intently on a single thought, she will begin to feel a "slight quiver" in the pineal gland, the result of a magnetic current flowing through the dense molecules of glandular tissue. "If the thought be strong enough to cause the current," Besant instructs, "then the thinker knows that he has been successful in bringing his thought to a pointedness and a strength which render it capable of transmission." Of course, successful thought transmission requires that the pineal gland of the recipient be equally evolved; for the uninitiated or the simply dull-witted, the loftier thought vibrations will fall, so to speak, on deaf ears. The adept could pick up thought vibrations not only from fellow mortals, but also from the spirits of those long dead (the lucky few could tap into the thoughts of the Ascended Masters, including Buddha and Jesus). Krotona's state-of-the-art meditation room was uniquely designed to facilitate the flow of cosmic signals of the most elevated variety, thereby eliminating any external impediments to the spiritual evolution of its members.

While Blavatsky and Besant's writings briefly touch on the Hindu concept of a chakra (*wheel* or *circle* in Sanskrit) as a vibrating center of bodily energy, the idea was much more vigorously taken up and popularized in the American context by Charles Leadbeater in his 1927 monograph, *The Chakras*. The seven chakra points are seven perpetually spinning energy vortices, located along the spine from sacrum to crown like flowers shooting forth from a central stem, and into which "a stream of force from the higher world is always flowing." Only one who has mastered the art of clairvoyance can see these otherwise invisible orbs, and they vary in shape and size depending on the refinement of the chakra-holder. In the "ordinary man," Leadbeater instructs us, the chakras appear as "small circles about two inches in diameter, glowing dully." In the fully enlightened being, however, they manifest as "blazing, coruscating whirlpools, much increased in size, and resembling miniature suns," each with its signature color-aura. Leadbeater helpfully provides full-color illustrations of his visions so that the reader might recognize a chakra (and determine its degree of health) if and when they ever chance to see one.

But the final spiritual fusion of healing vibrations, chakras, and crystals that has become an article of faith in contemporary LA's spa and wellness culture was a relatively late-stage addition to the original theosophical brew. From Leadbeater's color-coded guide to the chakras it was just one step to assigning appropriately colored crystals to each of the body's seven energy points. Joy Gardner's 1988 *Color and Crystals: A Journey through the Chakras* did just this, throwing in a theosophically derived material explanation for crystal healing as well. "During meditation," Gardner writes, "the silica particles in our heads become magnetically charged and align themselves around the pituitary and pineal glands, which stimulates [them]." "The human body is like a liquid crystal," she goes on to clarify, and the silica inside our bodies responds to the same molecular makeup in quartz, aligning us with universal energy flows.

Old school Theosophy's basic tenets are alive and well in twenty-first century Los Angeles, where crystal workers generally continue to ascribe therapeutic effects to vibrational sympathy between stone and human organism. Like a tuning fork that sets adjacent metal humming to the same frequency, the crystal's stable, harmonious arrangement of atoms, they reason, restores order to the more disorganized molecular mess of the human body. The appropriate crystal brought within proximity to our bodies regularizes its rhythms, gets its out-of-sync vibrations back into elemental harmony.

The aptly named Crystal Vibrations on the boardwalk in Venice Beach is more crystal supply shop than boutique, a no-nonsense emporium with crate after open crate of crystal, polished and raw, sold by weight. An entire glass case in the back of the shop is devoted to rose quartz. Yusef, a thin, polite young man with a jet-black ponytail, hands me a well-thumbed copy of Robert Simmons's *The Pocket Book of Stones* (which, I note, is the guidebook of choice among the several open-air crystal shops along the boardwalk). Rose quartz, the guide instructs me, is "the pure stone of love," and meditating with it "brings an envelope of love around oneself and activates the heart chakra." Nor is this all. Rose quartz vibrations "calm and cleanse the entire auric field," and can even "penetrate to the cellular level, reprogramming the cells for joy and longevity rather than despair and death." I ask Yusef about the difference between raw and polished quartz and why some stones are shaped to a point. "It's more a question of personal preference," he shrugs. "Some people don't like their stones to have been passed through a machine; they want them in their natural state." The pointed shafts, with their columnar striving toward the sky, are more effective than an oval or disk-shaped stone at "activating the crown chakra" during meditation, he explains.

All the quartz pieces at Crystal Vibrations, both polished and raw, are beautiful. They vary in hue from the deep blushing pink of a magnolia blossom to the barely there tint of watermelon ice. I choose a piece of the palest pink raw quartz, the size of a cherry, that fits easily in my closed palm. Yusef weighs it out, and I depart with my purchase.

Up on Main Street, as honky-tonk Venice transitions by imperceptible degrees into the more buttoned-down Santa Monica, the crystal shops are more curated. The wooden storefront of one aesthetically pleasing boutique is painted a soothing shade of lettuce green, the interior airy and light filled. Double glass doors at the back open onto a brick terrace studded with potted plants and giant amethyst geodes. The shop serves complimentary crystal-infused water from a glass samovar and sells a variety of crystal-adjacent aids to healing: aromatherapeutic incense bundles, herbal teas with crystal-studded tea ball strainers, an assortment of crystal earrings and bracelets. I am attracted to a pair of stud earrings featuring aquamarine crystal that claim to facilitate cosmic correspondence with souls from the lost city of Atlantis.

For ten dollars, I register for a "manifestation" meditation workshop hosted after hours on the patio. There are just three of us, and the meditation leader is a serene-looking blonde woman with an immaculate French-tip manicure and a white embroidered caftan, which is about what I expected when I signed up. She explains to us that we are all on a journey through tiered levels of consciousness, each one bringing us in closer touch with "the inner divine." The first level is characterized by passive victimhood: "life happens to me." As we ascend to second-level consciousness, the positions flip, and the self becomes an active agent: "I happen to life." The best of all states is third-level consciousness, our guide explains, where, with a slight shift of preposition, "life happens *through* me." We recognize ourselves as cocreator of what happens with "the source energy" moving through us. This stage requires concentrated listening, making our consciousness "available" to receive whatever the source energy is looking to manifest through us. "Channeling consciousness is not a forceful energy," she reminds us. "It is an energy of 'letting be'; que será, será."

I ask what part crystals play in ascending the scale. "For me, crystals represent the specific energy of the time and place deep in the earth where they came into being," she explains. "When you touch them or carry them with you, they remind you that you carry a piece of this universal energy within yourself." She recommends a rare species of stone, only recently discovered, called Auralite 23. Belonging to the amethyst family, it contains an astounding twenty-three minerals and is more than a billion years old. Through meditation, she says, we gradually absorb the crystal's energy until we reach channeling consciousness and no longer need the outside support of the physical stone. "At that point you can give the crystal to someone else who might need it on their journey," she recommends. I leave the class thinking crystal healing, at least of the Santa Monica variety, is like a genial DIY course in pop-Stoic philosophy and/or watered-down Buddhist meditation, with a side of mineral pseudoscience thrown in.

In the Western imagination, science and magic are often seen as opposing worldviews, but in fact they are animated by a similar impulse: our anguished animal desire to exercise control over the environment, to pull order out of chaos, and thereby (originally, at least) to boost our species' chances of survival. Science has developed a series of tests to confirm the existence of cause-and-effect connections; magic has not, and continues to place faith in causal relationships that don't stand up under scientific scrutiny. The Enlightenment sought to stamp out what it viewed as the primitive animism of an earlier stage of human evolution, but belief in a network of secret sympathies structuring the cosmos has proven remarkably resilient. And for good reason: according to anthropologist Bronislaw Malinowski, magic first emerged as an evolutionary stopgap when more empirically sound predictive models failed to guarantee human thriving. Magic's true purpose is to "ritualize man's optimism," Malinowski conjectures, and in this function represents the "sublime folly of hope"—a hope as groundless as it is necessary.

I am not unsympathetic to this long-standing human impulse to see, in stones, a kind of steadying talisman, a comforting messenger from the object world. Robert Simmons, patron saint of crystal shops, describes the feeling of "secret connection" that floods him upon first meeting a new stone, wondering "where it came from, how long had it waited underground to be discovered and brought to my hands." He, like my crystal shop meditation leader, credits stones with drawing humans outside the narrow confines of the self, deepening our awareness of all matter's "partnered destiny."

Among the more compelling explanations for the healing effects of stones that I happened to stumble upon in my research on crystals appears in W. T. Fernie's 1907 handbook *Precious Stones: For Curative Wear, and Other Remedial Uses*. Dr. Fernie suggests the geological ancientness of stones is in and of itself a possible tonic for what ails us. "Gems and Precious Stones retain within themselves a faithful and accurate record, even to its smallest detail, of physical conditions, and acquired properties, from the primitive time of their original molecular beginning," the author muses. Brought into contact with the body, the stone will "continue to assert meantime its long-remembered virtues." It should come as no surprise, then, that a well-chosen diamond, whose "adamantine brilliancy" represents "the sublimated sunshine of many tropical eons," might recharge the depleted spirit of its wearer.

I look at the rose quartz I hold in the palm of my hand, feel its heft, see how the light passes through it. In the end, though, I'm more attracted to a collection of flat beach stones I once picked up on the Greek island of Ios. In shades of pale gray, white, and pink, these stones are surf-tumbled smooth and almost eerily, perfectly circular, as though prefiguring the flat white faces of the fertility goddess statues the Minoans sculpted from island marble over five thousand years ago. I select an off-white stone about the size and thinness of a dime and rub its smooth surface between my thumb and forefinger.

To the extent that meditating on the molecular time scale of crystals relative to our own frail flesh might bring a kind of deep time wisdom, then, it is perhaps not impossible that the antique soul of a rose quartz or a diamond or an Auralite 23 or even a nondescript beach stone from the shores of a Greek island might bring peace to the human soul. Just not, or not entirely, for the reasons crystal healers suspect.

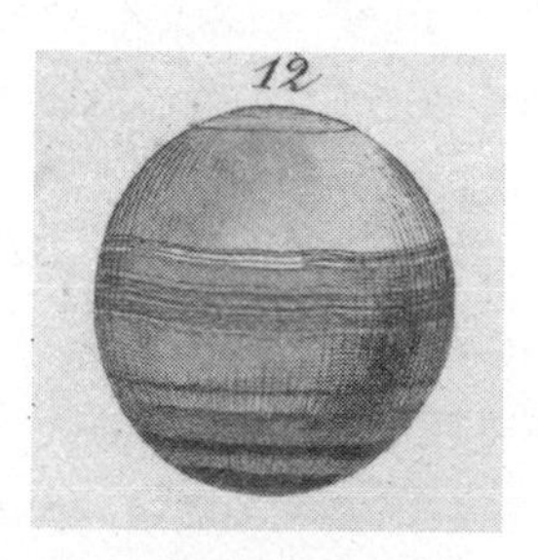

OBJECT PERMANENCE

I am looking at a picture of my father and me from when I was maybe three years old. His dark-brown hair is trimmed close, his beard neat, a precursor to the more shaggy, salt-and-pepper style that marks photos of him from the mid-'70s. He's wearing a cerulean blue sweater; I'm seated on his lap, with uneven pigtails and a red cotton dress and a mouthful of impossibly tiny white teeth. Our heads are bent together over a book he holds in lean, brown hands. Everything from the faded Kodachrome color palette to the yellowed edges of the print acts as a time stamp, assigning this moment to its place in a vanished past.

It is the anniversary of my father's death, four years gone, and I wonder today, as I do every day, *Where are you?* I mean the question quite literally. On what map, by what transmogrified geography of spirit and sinew, can I find you? Where is the thing that casts a shadow? "It doesn't feel real," we say after the death of a loved one, and indeed it does not. That is because death forces us to reverse the cognitive work of earliest childhood, one of the first-learned lessons that ushers us into the world of reality and reality-testing: object permanence. Every cognitively mature adult accepts this principle, more or less: just because an object is no longer in our sight, within reach, doesn't mean it has ceased to exist. Babies, of course, don't know this; once an object passes out of their field of vision, even a desired object like the mother's face, they assume it is gone for good. Once the infant learns that the mother still exists even when it can't see her, however, the way is paved for symbolic thinking: the infant creates an image of the mother in memory that they can hold onto as guarantor of her eventual return.

It is why babies go wild for peekaboo. When they cover their eyes with their hands, the mother's cherished face disappears for a moment and then, with a flick of the wrist, reappears: it has been there all along, the child realizes. Transitional objects—blankie, binkie, teddy bear—help manage the anxiety of separation, convince the child it still has ahold of the mother, if only by proxy. Once the mental image of the mother is impressed deeply enough upon the infant's memory to guarantee her permanence, a pledge of her return, then the more literal material supports can fall away.

Yet that state of innocence and play, loss and recapture, stays with us well beyond infancy. Magic tricks work by triggering archaic reality-testing questions we thought we had solved once and for all. We feel a whoosh of spooky surprise when the ball disappears beneath the cup, the scarf vanishes inside the magician's fist. Similarly, when a loved one dies, we can't stop searching for the rabbit inside the hat. "They have to be somewhere," we say as if we were searching for lost keys, a lost book, a missing earring. "They can't have just disappeared into thin air." And so, we are stumped by this grimmest of magic tricks: now you see it, now you don't.

There is a black-and-white photograph of my father as a young child, from the late 1930s, when he is maybe four—probably the same age as I am in the picture of us reading a book together. In this photo he is standing on the sidewalk outside of my

grandparents' house, bundled up in a buttoned coat and leggings and galoshes. The light is dull, wintry, sullen. He is holding a doll and staring at the camera with a quizzical, lightly sad expression: his eyebrows raised, a slight tug in his lip as though the camera operator were trying—and largely failing—to elicit a smile. What I'm drawn to in the photograph are his eyes: they are the same soft, brown, gentle eyes that I know not only from my own memories of him, but that I see in pictures of my sister when she was a child, and now in her children. This gene is like a life raft that carries a piece of my father away from the wreckage, brings him safely to the shore of the (still) living.

A photograph is made when light reflected from the surface of an object travels through the camera lens and hits the silver salt coating of the film; the photons catalyze a chemical reaction in the crystals. Wherever a photon lands, the crystal turns into to a speck of silver. The reflected light etches an image on the plate, which is why early photography was called heliography: sun writing. Had there been no energy expenditure—no beam of light triggering a chemical reaction in the silver salt—there would be today no trace of my father's soft, uncertain half-smile in the photograph I hold in my hand.

Memory works in an analogous way. Your neurons fire to gather the sensory stimuli you receive into predictable composites: the lineaments of your father's face; the grain of his voice, his comforting gaze and embrace, your head now (as an adult) fitting neatly into the pocket between his chin and chest when you hug him, cheek against his soft wool sweater. These experiences etch into your cells, burn fixed molecular pathways that allow you to recall him at will, within the camera obscura of memory, and predict his future reappearance. Here, too, work is done, heat released: your neurons change form, exchange chemical signals, link into chains to engrave these external objects on the gray matter, on the heart—not unlike the chemical transformations of the photographic plate. The more time you spend in the loved one's presence, the more you bind yourself to that person—not just in metaphor, but in molecules. To say you hold the loved one within your body is not poetic license. It is a fact of physics and biology.

Mourning a loved one, according to Freud, is a slow process of undoing object permanence, a staged series of lessons in letting go. The object is tied to the ego and "reinforced by a thousand links," he lectures, each one of which must be eventually broken. "Reality testing has shown that the loved object no longer exists," Freud writes of the normal mourning process, "and it proceeds to demand that all libido shall be withdrawn from its attachment to that object." This demand to renounce the object meets with obvious opposition, and reality's marching orders "are carried out bit by bit, at great expense of time and cathectic energy, and in the meantime the existence of the lost object is psychically prolonged." We gradually learn to withdraw our emotional attachment to the object, to "abolish" it in some sense by canceling our affective investment in all "memories and situations of expectancy" the object held for us. This is, from an economic point of view, a wasteful process. "Why this compromise by which the command of reality is carried out piecemeal should be so extraordinarily painful," Freud frets, "is not at all easy to explain in terms of economics."

After my father died, for a long time I would see him every night when I shut my eyes to go to sleep. The minute my lids closed his face would appear again, floating, just as it looked the seconds before he drew his last breath. His moment of passage is one of the clearest memories I will likely ever have. Each night when I saw him my throat would tighten. It was grief, but of a welcome sort: I knew each night I would see him and feel him once again with a vividness that had the force of a visitation.

Then, bit by bit, the visits stopped. After a year, his image would still loom up at intervals, but not predictably. It faded quickly, like an afterimage on the retina fades, the edges blurring and frayed until it disappears and leaves you looking out on dull reality: a pale shadow of the sun's once-bursting light.

Melancholics never forget, never break faith, which is why Freud distrusted them: they cheat the efficient calculus of grief. In order to hold on to the lost loved one, the melancholic transfers her investment in the outside object back onto the ego itself. All the rage and pain she feels at having lost the loved one is, instead, turned inward. There the ego "debases and rages against itself," lapsing into the blackest and most intractable of melancholy funks. But the logic behind this bait and switch is clear: "by taking flight into the ego, love escapes extinction," Freud reasons. Melancholics refuse the cold chemical stability of forgetfulness that allows all extinct cathectic bonds to be broken, their energy dispersed. Rather, the psychochemical dynamics of melancholy take all the energy that once bound the grieving person to her lost object and conserve it, albeit in disguised form. Melancholy is a refusal to accept the dissipation of energy and matter, a bulwark against the second law of thermodynamics: No, this object will not disappear, will not dissipate. For I've hidden it inside me like a flame, where no breath of outside air can blow it out.

My father taught me how to build a proper fire. Scrape the ashes out of the hearth; lay down two split logs side by side, then crisscross another two on top to form a hashtag. Stuff twists of newspaper and spindly sticks in the crevasses. "But not too tight, or you'll suffocate it," he warned. The flame needed oxygen to breathe. We would light a match, then lean back, brushing ash and splinters off our knees, to watch the pyre take flame: a liquid rim of orange and blue running along the grain of the wood. The newspaper ink was dull gray on his fingertips, but upon contact with the heat leapt up into wild tongues of green, like scrolls unfolding from the prophets' mouths in a medieval tapestry. I still see his face, lit by the red glow of the hearth.

Yet the physics of time tells us that object permanence of any sort is an illusion. Everything disappears in the end, and if our hearts are stuck on things reappearing this is only because of a perspectival error: we lack the long view by which the cosmos is ceaselessly unraveling itself. From the point of view of the universe, there is only flux, and clusters of matter that occasionally cohere, stubbornly hold together for a few brief moments in the cosmic drift toward total stasis. Time is just one way we can count the expenditure of energy, the gradual process by which brief islands of tottering homeostasis—my father, me, the earth, the sun—slowly (measured by what clock?) dissolve back into the cosmic current.

All memories are heat traces, and the future is heat that has not yet been kindled or spent. Another way to think of time's arrow is that it is a movement from the improbable toward the probable. All things being equal in the universe, the most probable state is the equilibrium of complete entropy, the dissolution of all particular forms, all energy spent. That means anything holding out against this eventual state of blank uniformity—life itself, its myriad humped and huddled forms feeding and fueling and reproducing for a season—is a flash in the pan, a mere vehicle through which the universe slowly but inexorably carries out its larger project: disintegration. The future is a chemical reaction, a transformation of energy, a heat trace that hasn't happened yet. A piece of marble, two hundred million years old, is about as substantial and enduring a thing as one can imagine. But it is still not, strictly speaking, an object—let alone a "permanent object"—when regarded cosmologically. Stones are, in the words of physicist Carlo Rovelli, better thought of as "events that for a while are monotonous." My father's life was a slice of cosmic improbability, a glimmer of an event that my own bundle of molecules had the random luck to be entwined with, for a season.

I don't know if this happened or not; I may have dreamed it. We used to go sledding on the big hill behind the golf course near my grandmother's house. We had plastic sleds but also a wooden toboggan (a word I always found annoying and old-fashioned when my grandmother used it; why not just call it a sled?) with turquoise waterproof padded seats and a thick, slicked-over rope you took in your two hands to steer its course. In the stand of trees at the bottom of the hill there was a thin metal waterspout poking up through the snow, where even in winter you could take a drink. The basin of the hill sloped down into the woods forming a slight clearing, filled with felled trees and brush that one time we raked together and lit on fire, gathered around the warmth in the middle of our sledding, seated on logs, an arc of blue sky cut out against the black branches above us. My father kneeling, his beard lightly crusted in ice, his nose reddened by the cold, fanning the fire. My father, the fire maker: kindling magic, conjuring warmth out of nothing. We held our mittened hands above the crack and sway of the fire until the tiny knots of ice stuck to our wool mittens had melted. Then our hands were left encased in soggy mitts, steamy warmth turning to bitter cold in measure as we moved further away from the fire circle, trudging home through the drifts.

Now when my father visits me in my dreams, he is younger. I see the father of my childhood: lithe and tan, his dark hair not yet flecked with gray. On hot summer days, he would take my sister and me to the neighborhood pool. I watched him dive into the deep end and swim underwater to where I waited in the shallow, his yellow-and-green checked swim trunks winking like an exotic fish beneath the wobbly crystal surface until he came up splashing and blowing, water running off his beard. I see him behind the desk in his office, smelling of crumbling books and chalk, and remember how he'd take me down the hall to his secretary's office for snacks: microwave popcorn, peppermints in a basket on her desk.

The further I travel away from him in arrow time, borne forward into the future and away from our last moment of contact, the more archaic are the images thrown up by my unconscious as the bridge on which we meet. As if, as I careen toward old age, he moves backward in time, becomes younger, like a character in a fairy tale. Perhaps one day in the future, I will be able to conjure him only as a child: see the quizzical eyebrow and soft brown eyes of a toddler circa 1938, braced for fall in thick zippered leggings and galoshes, holding a doll, gazing shyly at his father behind the camera lens. A world vanished but whose ancient particle traces subsist: the energy of the sun transferred to silver nitrate transferred to the firing synapses of my brain, still flickering in my skull on the edges of sleep.

The picture is a promise, written in the incontrovertible language of interstellar molecules, of energy burnt and transmogrified, dissipated but never truly lost. *I remember you*, it says. *I am still with you, flesh of your flesh.*

CAMERA OBSCURA

We parked in a dusty lot off the hill town of Kástro, a village on the Greek island of Sifnos. We were there to see the Church of the Seven Martyrs, a little chapel sited on a spit of land in the middle of the Aegean. We wound our way down the cliffside along a stone path that led out onto the promontory, wind whipping our hair and clothes. For some reason I kept turning over in my mind the line from T. S. Eliot's "The Waste Land" about the chapel perilous, "the empty chapel, only the wind's home"—the castle where, in Thomas Malory's rendition of Arthurian legend, Sir Lancelot almost succumbs to a witch. The Seven Martyrs' remote position out on a rock, the green waves tossing and breaking beneath it, made me think of it as a margin at the end of the world, and lonely and perilous—if beautiful—for that reason.

This was not the first time I'd visited the Church of the Seven Martyrs. My husband and I had traveled to Sifnos in the late 1990s when we were graduate students many, many years earlier, and now here we were again, this time with our children in tow. As we approached the chapel, my husband took my hand with the excitement of this return to a shared past: "Remember," he prodded me, "how amazed we were the first time we saw this?" We'd rented a moped, apparently, leaving it in town to trace the very path I was now retreading out to the windswept chapel.

But, in fact, I had no memory of that earlier visit. Nothing in the landscape, nothing in the air or light, triggered a sense of having been here before. The warm yellow light of the setting sun bounced off the white church and blazed its signature on my retina as if for the first time.

This failure of recall didn't really surprise me; I have a terrible memory, a fact that anyone who knows me will readily confirm. I don't mean I have a bad memory for names and dates, the silly tiny stuff (although I am bad enough at that too), but for those essential blocks of experience with which normal people construct a life: the geography of cities I've lived in, places I've visited, the faces of people I've known, milestones in the lives of my children. Large swathes of life-stuff that most people's sensory apparatus naturally transcribes—catalogue under Things-to-Remember and store for later retrieval—go missing in my case. It has always been a source of great shame to me and something I try to keep hidden.

It is as though the usual thickness of points of contact with the world outside one's skin that give the average person a sense of grounding as they move through life have been thinned in my case: sporadic, tenuous, full of inexplicable gaps—as though my body and brain are but minimally inscribed in the space-time continuum.

It's not that I remember nothing, of course. I remember *another* trip to the Greek islands my husband and I made when we were in our twenties, this time to Amorgos. We visited a monastery there, I recall, perched like a chalk-white barnacle on the side of a cliff over the sea. As we made our way up to the entrance, we spied a priest in cassock and pillbox *skufia* bending over some piece of work in the shade of a tree. On our approach, he turned toward us, smiling through his gray beard, and we saw he held a knife in one hand and in the other a freshly killed, bloody bird that looked like a dove. Placing the bird and knife tenderly on a table, he wiped his hands on his cassock and ushered us inside. As a parting gift he offered us each a piece of Turkish delight, arrayed like dusty jewels in a burnished wooden box. I remember that. But of the Church of the Seven Martyrs: nothing, not a trace.

Such memory blanks are puzzling enough for me, but even more so for those close to me, who note this absentee quality in my mode of being and occasionally, rightfully, feel derailed by it, almost betrayed. I am both here and elsewhere, present and absent: now I see you, now I don't.

I didn't tell my husband that I retained no memory of our earlier Sifnos trip. I knew it would make him feel lonely, like he'd been living with a ghost.

Philosophers have always been attracted to the image of light streaming into a dark chamber and projecting pictures on the wall as a metaphor for how the mind, and in turn, memory, works. For Plato, the mind was like the smooth interior surface of a cave where objects, paraded before a flame, cast their images in a dusky outline like shadow puppets. Almost two millennia later, the English philosopher John Locke compared the mind to a camera obscura, early forerunner of the pinhole camera: "For methinks the understanding is not much unlike a closet wholly shut from light," writes Locke in his 1689 *Essay Concerning Human Understanding*, "with only some little opening

left, to let in external visible resemblances, or ideas of things without."

This philosophical preoccupation with light is natural enough when we consider that reacting to sunlight—sensing it, moving toward it, making pictures out of it—was among the earliest tricks biological organisms had to learn if they hoped to survive in the world at all. Cyanobacteria, the biological ancestors of all plants and animals, burst onto the primeval scene 2.7 billion years ago with their ability to turn sunlight into energy, and photoreceptor proteins in their single-celled bodies allowed them to track and then scoot their way toward the light. Early unicellular creatures conserved these photosensitive proteins and added pigments that, by deflecting certain of the sun's wavelengths, helped them track the light's source. By a series of chance mutations, these patches of photoreceptor proteins in early protists floating around in primordial soup slowly evolved into divots, then into cups, then into tiny little protoplasmic chambers with just one small opening for light entry—a cellular pinhole camera, in effect—which first allowed image capture. And from there, it was a gradual process of sharpening up the image.

Yet to trace the evolution of the human eye is to suspect nature of being a genius with a slightly dark sense of humor, so haphazard is the trajectory by which our sight organs came into being. When a random mutation in some ancestor metazoan chanced to pick up a cell-repair molecule called a crystallin and plop it, willy-nilly, onto the surface of a primitive eye—like blindly moving around parts on a Mr. Potato Head—the crystallin's molecular structure chanced "to have the right optical properties for bending light," and was duly pressed into service to fine-tune the images the creature could capture. What zoologists call the "Cambrian explosion," a dramatic flowering of animal diversity between 545 and 530 million years ago, was likely catalyzed by an increase in eye complexity. One scientist called it an evolutionary "arms race" to capture and focus light. And when our proverbial fishy ancestors made their first forays up onto land somewhere around 385 million years ago, the eyes again led the way: eyes tripled in size and migrated from the sides to the top of the head long before fish traded in fins for limbs. Early tetrapods, scientists suppose, cruised around in the sandy shallows, eyes popped up above the surface crocodile-style, preying on insects and millipedes at the water's edge and daring themselves, eventually, to make the leap to dry land.

Indeed, eyes appear in all sorts of unexpected places, and evolution will cobble together whatever is at hand to funnel light and grab an image. A humble mollusk called a chiton navigates the ocean floor sheathed in a coat of limestone armor studded with hundreds of tiny, mineralized eyes, a protoplasmic brooch. Over time, small genetic tweaks in the aragonite minerals composing the creature's shell—bigger grains here, a more orderly structuring of crystal molecules there—reduced the light-scattering properties of the "non-seeing armor" enough to capture a primitive image when light passed through these patches. Looking at a fish-shaped silhouette through an aragonite lens, researchers glimpsed "a somewhat blurred, but recognizable, shape." To dip into this history of sight is to sense the whole earth buzzing, scanning, probing, watchful: all life, hungry for light.

~

On Sifnos, we stayed in the tiny, sun-soaked, bay-bound hamlet of Vathi, where the water was smooth and shallow and blue for yards and yards. The day after our visit to the Church of the Seven Martyrs, I snorkeled across the bay, moved as always by the softness of the underwater light: silent, hazy yellow columns stippling the white sand seabed as if in slow motion, as though it had traveled eons to get there and had lost something of itself along the way, muted and tired, somehow, after its long journey (which was not, of course, only a metaphor; this spectral dimming is what *actually* happens to light as it travels down through water). As I lazily propelled myself deeper out toward the mouth of the bay, the element shifted from washed-out petal green to the color of sea-smooth bottle glass, to murky jade. It struck me suddenly that this was the way my brain had felt that day on the chapel rock: my memory's images watery and wobbly, thin, close to dissolution, shading off to blank or black. How dreary to be such a poor keeper of the light! To live shut up in this darkened closet, a camera obscura whose aperture regularly dilutes the sun's bright rays or, just as often, snaps shut against them altogether.

And yet other times I think: So what if I am a shoddy light-collector? So what if much of the light that falls on me bounces off and scatters to the far reaches of the universe? That I am granted a lens on the world at all—unwitting inheritor of ancient genes pressed into hard-scrabble, serendipitous service some five hundred million years ago, jerry-rigging up an underwater eye—is a minor miracle, no matter how foggy and tarnished my own model might be.

What I sometimes call my absenteeism—my sensing mechanism's persistent opacity, its determination to squander so much of the light that hits it—could be seen, not as a glitch in the system, but the price to be paid for being a body at all, a self opened out on a world that is not-self. It is only because we are *not* transparent mediums—rather, glass dark and earth-bound—that we can capture an image of the world in the first place. In Nathaniel Hawthorne's story "The Birthmark," the alchemist Aylmer's wife Georgiana is the very portrait of ethereal beauty save for a tiny, hand-shaped birthmark on her cheek. Aylmer becomes obsessed with the "stain on the poor wife's cheek," a visible token of the "ineludible gripe in which mortality clutches the highest and purest of earthly mould, degrading them into kindred with the lowest, and even with the very brutes." He repairs to his laboratory to concoct an elixir to remove it, and Georgiana dutifully drinks it down. As the last trace of the crimson hand fades from the lady's white cheek, so "the parting breath of the now perfect woman passed into the atmosphere, and her soul, lingering a moment near her husband, took its heavenward flight." Sin-haunted descendant of the Puritans as he was, Hawthorne knew that the realm of pure light, that unmixed with earth, is the sole province of God.

The body provides its own Hawthorne-style allegorical lessons. Take vitreous floaters, those strands of protein that get trapped in the gel inside your eye and cast shadows on the retina, so that when you look at the sky or a blank wall you see bobbing blobs like crystalline chromosomes, vermicelli semicolons, hopping

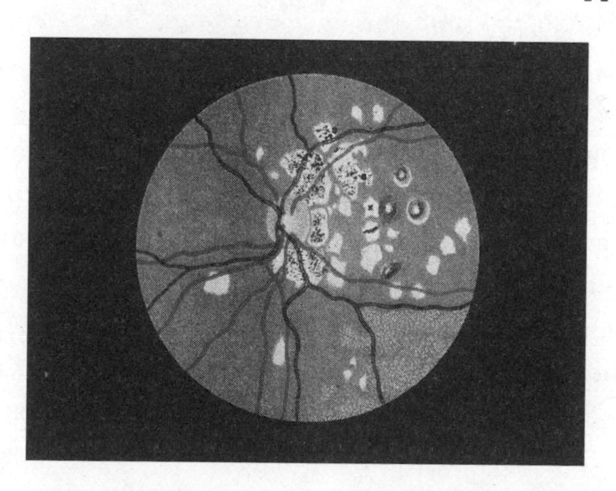

across your eyeballs. This is your body—its coagulating proteins, its material bunchings—getting in its own way, interposing itself like a dark planet between your eye and the sun of the world. These minieclipses are a salutary reminder of our embodied state, that we have access to these images—this stream of stuff coming in from the outside, light inscribed on our retina—only because our bodies provide miraculous portals, always subject to obfuscation or closure, onto a world beyond the black box of the self.

Fine, I think, as I track these funny little darkling swarms across the surface of my vision: I'll take what I can get.

When it comes down to it, the light I've lost was never mine to begin with, light-gathering being always a shared project. In the Judeo-Christian tradition, eschatology—from the Greek *eschatos*, "last, remote, uttermost, furthest"—is that branch of theology devoted to the science of last things: signs by which to recognize that the world is drawing to a close before its final redemption. The tradition of the *shevirat ha-kelim* in Jewish mysticism recounts the story of creation and redemption this way: on the first day, God contracted himself to make room for the world. He poured his light into ten vessels, which then shattered, scattering the divine ray into fragments. This fragmentation is the fallenness of earth, and our task is to gather the scattered sparks to restore the world. As the time of last things approaches, "when the task of gathering the sparks nears completion," Rabbi Menachem Mendel of Rimanov glossed, "God will hasten the arrival of the final redemption by Himself collecting what remains of the holy sparks that went astray." Viewed from a geo-biological-cosmic perspective, the wisdom of the kabbalah checks out: all life on earth is just (*just!*) the extended and highly improbable unfolding, through time and space, of an initial burst of solar grace; gift of light and heat from a burning, spinning star, which we receive and share and recycle endlessly, until the universe stops moving.

The Manichean Gnostic tradition has a similar eschatology, a tale of scattered sparks and their final collection. At the end of time, runs one Manichean psalm, "All Life, the Remnants of Light in every Place / [God] shall gather to himself and form of it an Image (*Eidolon*)." It is a beautiful myth, and I choose to picture it like this:

Wading in the shallows and gazing out on the bay at Vathi, sometimes I was able to make out a passing school of fish by a sudden flash of light beneath the water's surface, an electric leap that disappeared as quickly as it came. Schooling fish mold their individual bodies into incredible collective shapes—spiral, sphere, plane—coordinating with each other by mechanisms that scientists don't yet fully understand. The fish have special crystals beneath their scales that reflect the sunlight, and each must turn just so to catch and propel the light along their collective body in a uniform wave, weaving skin and nerve and bone and crystal together into an improbable emblem of light, an aquatic chorus line. Viewed from above, it must look like a giant silver orb winking at the sky from beneath the salt waves before, once again, closing shut, disappearing into the deep, a thousand tiny lights quenched by darkness.

I try to make peace, then, with the dusky vessel that I am, make peace with the light that goes missing. In the interim, I gather up those scattered remnants of light that do chance across my vision: the curious gaze of a tiny opal sea bream, butting up against my snorkel mask; the shiny, bobbing blue underside of a fishing boat; the ribbon-flash of a green eel in the rock. I offer them up to the collective Image that—perhaps—some cosmic light-keeper is patiently building for the end of time.

NOTES

Preface

x **In Aesop's fable** *Aesop's Fables*, trans. Laura Gibbs, reprint ed. (Oxford University Press, 2008), 235.

x **For the Anishinaabe** Joe McLellan, *Nanabosho: How the Turtle Got Its Shell* (Winnipeg: Pemmican Publications, 2015).

x **stout lizard** "Key Link in Turtle Evolution discovered," Smithsonian Insider, June 25, 2025, accessed December 19, 2023, https://insider.si.edu/2015/06/key-link-in-turtle-evolution-discovered.

x **highly conserved** Josh Davis, "Turtles have lived for 230 million years—but will they survive climate change?" Natural History Museum, May 22, 2020, accessed December 19, 2023, https://www.nhm.ac.uk/discover/news/2020/may/turtles-230-million-years-will-they-survive-climate-change.html.

xi **Landcrab I** "Landcrab I," in Margaret Atwood, *Selected Poems II: Poems Selected and New, 1976–1968* (Boston: Houghton and Mifflin, 1987), 59.

xi **In an 1871 letter** Juli Peretó, Jeffrey L. Bada, and Antonio Lazcano, "Charles Darwin and the Origin of Life," in *Origins of Life and Evolution of Biospheres*, vol. 39 (2009), 395–406, https://doi.org/10.1007/s11084-009-9172-7.

xi **life first took shape** Rachel Brazil, "Hydrothermal Vents and the Origins of Life," *Chemistry World*, April 16, 2017, accessed December 19, 2023, https://www.chemistryworld

.com/features/hydrothermal-vents-and-the-origins-of-life/3007088.article.

xiv **David Brower** John McPhee, *Basin and Range*, reissue ed. (New York: FSG, 1982), 135.

Gravity

4 ***[Laughter] is one of the clearest marks*** Charles Baudelaire, "De l'essence du rire," in *Oeuvres complétes*, vol. 2, (Editions Gallimard, 1976), 530 (my translation).

4 ***How art thou fallen*** Isaiah 14:12–13, KJV.

4 ***Better to reign in hell*** John Milton, *Paradise Lost*, 1.263.

6 ***What has four legs in the morning*** Apollodorus, *The Library*, book III, chapter 5, section 8, https://www.theoi.com/Text/Apollodorus3.html.

8 ***slid around in his black gloves*** Mark Twain, *The Adventures of Huckleberry Finn* (New York: Harper & Brothers, 1885), 250.

8 ***the softest, glidingest, stealthiest*** Mark Twain, *Huckleberry Finn*, 250.

11 ***labor does you*** Maggie Nelson, *The Argonauts* (Minneapolis: Graywolf, 2015), 134.

11 ***Pain is the mother's safety*** W. O. Priestley et al., *The Obstetric Memoirs and Contributions of James Y. Simpson, M.D. F.R.S.E.*, vol. 2 (Edinburgh: Adam and Charles Black, 1855–6), 608. https://hdl.handle.net/2027/ucm.5325486957.

12 ***Full fathom five*** William Shakespeare, *The Tempest*, act 1, scene 2.

12 ***There is something in us*** Flannery O'Connor, "The Grotesque in Southern Fiction," in *Mystery and Manners: Occasional Prose*, ed. Sally and Robert Fitzgerald (New York: Farrar, Straus and Giroux, 1969), 38.

13 ***Son of man*** Ezekiel 37:3, KJV.

13 ***there was a noise*** Ezekiel 37:7, KJV.

13 ***foregone optimism*** O'Connor, "Catholic Novelists and their Readers," in *Mystery and Manners,* 182.

Corpus Christi

16 ***Christ, who was still believed*** Miri Rubin, *Corpus Christi: The Eucharist in Late Medieval Culture* (Cambridge: Cambridge University Press, 1991), 68–69.

16 ***broken and chewed*** Rubin, *Corpus Christi*, 14.

17 ***savage mind*** James George Frazer, *The Golden Bough: A Study in Magic and Religion* (New York: Macmillan, 1926), 324.

18 ***Her lament is for the woods*** Frazer, *Golden Bough*, 326.

18 ***eating out of a drum*** Frazer, *Golden Bough*, 351.

19 ***The ecstatic frenzies*** Frazer, *Golden Bough*, 356.

19 ***the tide of Oriental invasion*** Frazer, *Golden Bough*, 357.

19 ***disguised under a decent veil*** Frazer, *Golden Bough*, 356.

19 ***When we call corn Ceres*** Preserved Smith, "Christian Theophagy: An Historical Sketch," *The Monist* vol. 28, no. 2 (April 1918), 161–208.

21 ***Verily I say unto you*** Matthew 25:40, KJV.

24 ***the food consumed*** Frederic L. Holmes, "The Intake-Output Method of Quantification in Physiology," *Historical Studies in the Physical and Biological Sciences*, vol. 17, no. 2 (1987), 235–70.

25 ***When I ask myself*** Flannery O'Connor, *The Habit of Being* (New York: Farrar, Straus, and Giroux, 1979), 92.

Outis

30 ***One nineteenth-century psychologist*** Pierre Janet, *Les Obsessions et la Psychasthénie* (Paris: Félix Alcan, 1919), 79, https://archive.org/details/lesobsessionsetl01jane, (my translation).

32 ***Robert Burton, whose 1621*** Robert Burton, *The Anatomy of Melancholy* (Philadelphia: E. Claxton and Co., 1883), 249, https://archive.org/details/anatomyofmelanch00burt.

33 ***In all other things*** Burton, *Anatomy of Melancholy*, 236.

33 ***'Tis not amiss to bore*** Burton, *Anatomy of Melancholy*, 408.

34 ***locked in impotence*** Henry Maudsley, *The Pathology of Mind: A Study of Its Distempers, Deformities, and Disorders* (London and New York: Macmillan and Co., 1895), 204, https://wellcomecollection.org/works/qjannfrc/items.

34 ***all the while perfectly conscious*** Maudsley, *Pathology of Mind*, 171.

34 ***come and stay there*** Maudsley, *Pathology of Mind*, 171.

34 ***reflex organic machine*** Maudsley, *Pathology of Mind*, 195.

35 ***Feeling of Incompletion*** Pierre Janet, *Les Obsessions*, 318.

35 ***I was listening*** Janet, *Les Obsessions*, 314.

37 ***The headless file*** Jean-Henri Fabre, *The Life of the Caterpillar*, trans. Alexander Teixeira de Mattos (New York: Dodd, Mead and Co., 1916), https://www.gutenberg.org/cache/epub/66762/pg66762-images.html.

37 ***The obsessive patient*** Janet, *Les Obsessions*, 157.

40 ***According to authors*** Kara Gavin, "Stuck in a Loop of 'Wrongness': Brain Study Shows Roots of OCD," *Michigan Medicine* (November 29, 2018), https://www.michiganmedicine.org/health-lab/stuck-loop-wrongness-brain-study-shows-roots-ocd.

41 ***bandaged moments*** Emily Dickinson, "The Soul Has Bandaged Moments" from *The Poems of Emily Dickinson: Variorum Edition* (Cambridge, MA: Harvard University Press, 1998), 1:385.

41 ***It was not Death*** Dickinson, 1:379–80.

42 ***There shall no man*** Exodus 33:20, 22, 23, KJV.

42 ***Standing on the bare ground*** Ralph Waldo Emerson, *Essays and Lectures*, Library of America, vol. 15 (New York: Penguin Putnam, 1983), 10.

Body Map

45 ***The principle of all things*** Quoted in S. K. Heninger, *Touches of Sweet Harmony: Pythagorean Cosmology and Renaissance Poetics* (San Marino: Henry E. Huntington Library and Gallery, 1974), 78.

47 ***I have been half in love*** John Keats, "Ode to a Nightingale," Poetry Foundation website, accessed December 11, 2023, https://www.poetryfoundation.org/poems/44479/ode-to-a-nightingale.

47 ***cease upon the midnight*** "Ode to a Nightingale."

48 ***Neural Maladies of Stumps*** S. Weir Mitchell, *Injuries of Nerves and Their Consequences* (Philadelphia: J. B. Lippincott and Co., 1872), 348, https://archive.org/details/injuriesofnerves00mitcuoft.

48 ***Nearly every man*** Mitchell, *Injuries of Nerves*, 349.

48 ***Until it [touched]*** Mitchell, *Injuries of Nerves*, 351.

48 ***We are competent*** Mitchell, *Injuries of Nerves*, 351.

49 ***Wilder Penfield*** Mo Costandi, "How sights, sounds, and touch are mapped onto the brain," *The Guardian*, September 3, 2013, https://www.theguardian.com/science/neurophilosophy/2013/sep/03/topographic-mapping.

49 ***The map is not the territory*** S. I. Hayakawa and Alan R. Hayakawa, *Language in Thought and Action*, fifth Harvest ed. (New York: Harcourt, 1990), 19.

50 ***Mitchell published anonymously*** Silas Weir Mitchell, "The Case of George Dedlow," *The Atlantic* (July 1866), https://www.theatlantic.com/magazine/archive/1866/07/the-case-of-george-dedlow/308771/.

50 ***I found to my horror*** Mitchell, "The Case of George Dedlow."

51 ***"Indeed," writes neuroscientist*** Vilayanur S. Ramachandran et al., "Phantom Limbs and Neural Plasticity," *Neurological Review*, vol. 57 (March 2000), 317–20.

53 ***By what evidence*** John Stuart Mill, *An Examination of Sir W. Hamilton's Philosophy and of the Principal Questions Discussed in His Writings*, 2nd ed. (London: Longmans, Green and Co., 1865), 208.

53 ***The infant uses*** Andrew N. Meltzoff and M. Keith Moore, "Imitation of Facial and Manual Gestures by Human Neonates," *Science*, New Series, vol. 198, no. 4312 (Oct. 7, 1977), 75–78.

54 ***1992 study*** Giuseppe Di Pellegrino et al., "Understanding motor events: a neurophysiological study," *Experimental Brain Research*, 91 (1992): 176–80, https://doi.org/10.1007/BF00230027.

54 ***the first- and third-person*** Vittorio Gallese, "Embodied Simulation: From Neurons to Phenomenal Experience," *Phenomenology and the Cognitive Sciences*, 4 (2005), 23–48, 38.

Turning, Unfolding, Passing Through

59 ***configuration and colour*** Aristotle, *On the Parts of Animals*, trans. W. Ogle (London: Kegan, Paul Trench and Co., 1882), 6, https://archive.org/details/aristotleonparts00arisrich.

59 ***virtue of what force*** Aristotle, *Parts of Animals*, 5.

59 ***For it is not enough*** Aristotle, *Parts of Animals*, 6.

60 ***To watch an animal egg*** Edmund W. Sinnott, "Biology and Teleology," *Bios*, vol. 23, no. 1 (March 1954), 35–43.

60 ***without form, and void*** Genesis 1:2, KJV.

67 ***Are not two sparrows*** Matthew 10:29–30, KJV.

67 ***great and small*** 1 Corinthians 15:54, KJV.

68 ***To any one who has ever*** William James, *Pragmatism: A New Name for Some Old Ways of Thinking* (New York: Longmans, Green and Co., 1907), 95, https://archive.org/details/157unkngoog.

American Pastoral

70 ***the grain-giving earth*** Hesiod, *Theogony, Works and Days, Testimonia*, ed. and trans. Glenn W. Most (Cambridge, MA: Loeb Classical Library, 2006), 97, https://archive.org/details/hesiod-hesiod.

71 ***fragrant reeds*** Theocritus, "The Harvest-Home," in *The Greek Bucolic Poets*, trans. J. M. Edmonds, 105, https://archive.org/details/L028GreekBucolicPoetsTheocritusBionMoschus.

72 ***On an occasion such as*** Nelson A. Rockefeller (governor of New York), from his speech at the dedication ceremony for the Nine Mile Point Nuclear Power Plant, October 9, 1969, 1:30 p.m., Oswego, New York.

73 ***Few things look as dead*** Stevenson Swanson, "Alewives Are Back, Dying on Beaches," *Chicago Tribune*, May 17, 1992, accessed December 12, 2023, https://www.chicagotribune.com/news/ct-xpm-1992-05-17-9202140150-story.html.

77 ***the peaceful power*** Rockefeller, speech.

78 ***You know, every time*** *The Beast from 20,000 Fathoms*, dir. Eugene Lourie, Warner Bros. Pictures, 1953, 5:44–6:19.

79 ***Do you know what*** *The Beast*, 5:44—6:19.

This Ragged Claw

84 ***mental distress or agitation*** Merriam-Webster.com Dictionary, s.v. "worry," accessed December 11, 2023, https://www.merriam-webster.com/dictionary/worry.

85 ***A score of 0*** Kiara Anthony, "BI-RADS Score," Healthline.com, September 29, 2018, accessed December 11, 2023, https://www.healthline.com/health/birads-score#scoring.

86 ***Cancer has always been*** Alanna Skuse, *Constructions of Cancer in Early Modern England: Ravenous Natures* (Palgrave Macmillan, 2015), 632–48.

87 ***verie hardly pulled awaie*** Skuse, *Constructions of Cancer,* 30.

87 ***slea the worme*** Skuse, 641.

87 ***prodigious Multitude*** Skuse, 642.

87 ***meat cure*** Skuse, 642.

88 ***the Wolf peeps out*** Skuse, 632.

88 ***like cures like*** My own definition of *homeopathy.*

88 ***wolf tongue*** Skuse, 644.

88 ***stamped and strained with Ale*** Skuse, 641.

90 ***O Rose thou art sick*** William Blake, "The Sick Rose" in *Songs of Innocence and Experience* (New York: Abrams, 2007), 39.

91 ***poor homeless wanderers*** Franz Kafka, "The Burrow" in *The Complete Stories* (New York: Knopf Doubleday, 1988), 327.

92 ***audible only to the ear*** Kafka, "The Burrow," 348.

92 ***but the noisemakers*** Kafka, "The Burrow," 348.

92 ***danger lies in ambush*** Kafka, "The Burrow," 352.

94 **waeccan** Online Etymology Dictionary, s.v. "wait," accessed December 11, 2023, https://www.etymonline.com/word/wait.

94 ***moving over the face*** Genesis 1:2, KJV.

Natural Magic

98 ***But the positioning*** Ptolemy, *Ptolemy's Tetrabiblos, or Quadripartite: Being Four Books of the Influence of the Stars,* trans. J. M. Ashmand (London: W. Foulsham and Co., 1900), 4, https://archive.org/details/ptolemystetrabib00ptol.

100 ***Robert Turner's 1664*** Robert Turner, *Botanologia: The Brittish Physician, or The Nature and Vertues of English*

Plants (London: Printed by R. Wood for Nath. Brook, 1664), 5, 35, https://wellcomecollection.org/works/eh4tzm96.

100 ***If you want your body*** Marsilio Ficino, *Three Books on Life* (Tempe, AZ: The Renaissance Society of America, 1998), 247.

101 ***The above-mentioned things*** Ficino, 249.

101 ***draw the power*** Ficino, 249.

101 ***circle of little animals*** My own translation from the Greek.

106 ***root of hemlock*** William Shakespeare, *Macbeth*, act IV, scene I.

106 ***Edward Topsell's 1658*** Edward Topsell, *The History of Four-Footed Beasts, Serpents, and Insects* (London: Printed by E. Coates, for G. Sawbridge, 1658), 198, 629, https://archive.org/details/historyoffourfoo00tops.

107 ***the Plants and Grass*** Turner, 7–8.

107 ***Biologists in the 1970s*** "A Story of Discovery: Natural Compound Helps Treat Breast and Ovarian Cancers," *The National Cancer Institute*, March 31, 2015, accessed December 11, 2023, https://www.cancer.gov/research/progress/discovery/taxol.

The Atavist

111 ***[God] has chosen*** Robert Chambers, *Vestiges of the Natural History of Creation* (New York: Wiley and Putnam, 1845), 177, 179, https://hdl.handle.net/2027/hvd.32044000262931.

111 ***The great Tree of Life*** Charles Darwin, *On the Origin of Species*, anniv. ed. (New York: Signet, 2009), 127.

112 ***the heart, for instance*** Chambers, *Vestiges*, 165, 164.

113 ***watery grey eyes*** H. G. Wells, *The Island of Dr. Moreau* (Garden City, New York: Garden City Publishing, 1896), 15.

113 ***dropping nether lip*** Wells, *Dr. Moreau*, 15.

113 ***misshapen man*** Wells, *Dr. Moreau*, 21.
113 ***forming something dimly*** Wells, *Dr. Moreau*, 21–22.
113 ***ocean menagerie*** Wells, *Dr. Moreau*, 24.
113 ***number of grisly staghounds*** Wells, *Dr. Moreau*, 23.
113 ***a huge puma*** Wells, *Dr. Moreau*, 23.
113 ***a solitary llama*** Wells, *Dr. Moreau*, 24.
114 ***to find out the extreme limit*** Wells, *Dr. Moreau*, 137.
114 ***animals half wrought*** Wells, *Dr. Moreau*, 245.
114 ***These tiny creatures*** Herbert Snow, *A General Theory of Cancer-Formation, being a Lecture Delivered at the Cancer Hospital on March 1st, 1889* (London: J. & A. Churchill, 1889), 20.
115 ***ancestral cellular functions*** P. C. W. Davies and C. H. Lineweaver, "Cancer Tumors as Metazoa 1.0: Tapping Genes of Ancient Ancestors," *Physiological Biology*, vol. 8, no. 1 (February 7, 2011), 2, http://dx.doi.org/10.1088/1478-3975/8/1/015001.
115 ***to up the mutation*** Mark Vincent, "Cancer: A De-Repression of a Default Survival Program Common to All Cells?," *BioEssays*, vol. 34, issue 1 (November 22, 2011), 72–82, 73, https://doi.org/10.1002/bies.201100049.
115 ***competitive struggle*** Vincent, "Cancer," 73.
117 ***abjure[d] all roofs*** William Shakespeare, *King Lear*, act 2, scene 4, lines 205, 207.

Quartz Contentment

119 ***Things will be fine*** Donni Saphire (@donni), Things will be fine, eventually, in thousands of years, for rocks, Twitter (X), April 7, 2020, https://twitter.com/donni/status/1247729433366183937.
120 ***One contemporary geological*** William W. Mather, *Elements of Geology for the Use of Schools and Academies*, 4th ed. (New York: Clement and Packard, 1841), 32, https://books.google.com/books?id=N4NJAAAAIAAJ.

121 ***Forehead copie[s] stone*** Emily Dickinson, *The Poems of Emily Dickinson*, Variorum ed. (Cambridge, MA: Harvard University Press, 1998), 2:608.

121 ***Safe in their Alabaster*** Dickinson, 1:159.

122 ***How happy is the little stone*** Dickinson, *The Complete Poems*, part 2, section 33, lines 1–4, 7.

125 ***Aristotle called a "nutritive" soul*** Susannah Gibson, *Animal, Vegetable, Mineral: How Eighteenth-Century Science Disrupted the Natural Order* (Oxford: Oxford University Press, 2015), 13.

126 ***Aristotle called a "rational soul"*** Gibson, *Animal, Vegetable Mineral*, 13.

126 ***Stones grow*** Gibson, *Animal, Vegetable Mineral*, 153.

126 ***Your letter*** Emily Dickinson, letter to brother Austin, February, 17, 1848.

127 ***And then I could not see*** Dickinson, 2:172.

127 ***Regardless grown*** Dickinson, 1:396.

127 ***But most, like Chaos*** Dickinson, 1:380.

128 ***without form, and void*** Genesis 1:2, KJV.

128 ***Vernadsky refers to the entirety*** Vladimir I. Vernadsky, *The Biosphere*, trans. David B. Langmuir (New York: Copernicus Springer-Verlag, 1998), 61.

129 ***Life begets rock*** Maya Wei-Haas, "Life and Rocks May Have Co-Evolved on Earth," *Smithsonian Magazine*, quoting Robert Hazen, January 13, 2016, https://www.smithsonianmag.com/science-nature/life-and-rocks-may-have-co-evolved-on-earth-180957807/.

129 ***On the organic side*** Lucas Joel, "How Did Animals Get Their Skeletons?: Siberian Rocks Provide Clues," Science.org, November 15, 2016, accessed December 11, 2023, https://www.science.org/content/article/how-did-animals-get-their-skeletons.

130 ***One generation passeth*** Ecclesiastes 1:4, KJV.

Lapidary Medicine

131 ***Knesko, whose line*** "Our Story," Knesko Skin, accessed August 13, 2023, https://knesko.com/pages/our-story.

134 ***The diamond, or what*** Pliny the Elder, "Adamas," in *The Natural History of Pliny*, trans. John Bostock and H. T. Riley (London: H.G. Bohn, 1857), 6:408.

134 ***no green in existence*** Pliny the Elder, "Smaragdus," in *Natural History of Pliny*, 6:409.

135 ***raise men's souls*** Joan Evans, *Magical Jewels of the Middle Ages and the Renaissance, Particularly in England* (Oxford: Oxford University Press, 1922), 76, https://hdl.handle.net/2027/mdp.39015076335093.

136 ***deserted sanctuaries*** H. P. Blavatsky, *Isis Unveiled*, vol. 1 (New York: J. W. Bouton, 1877), vi, https://hdl.handle.net/2027/hvd.32044023317761.

137 ***There is one word*** Annie Besant, *Thought Power: Its Control and Culture*, reprint (Wheaton, IL: Theosophical Press, 1953), 13, https://archive.org/details/thoughtpower.

137 ***If the thought be strong*** Besant, *Thought Power*, 33.

138 ***Krotona's state-of-the-art*** Hadley Meares, "The Creation of Beachwood Canyon's Theosophist 'Dreamland,'" *Curbed Los Angeles*, May 22, 2014, accessed December 11, 2023, https://la.curbed.com/2014/5/22/10099768/the-creation-of-beachwood-canyons-theosophist-dreamland.

138 ***In the "ordinary man*** C. W. Leadbeater, *The Chakras: A Monograph* (Chicago: Theosophical Press, 1927), 2, https://hdl.handle.net/2027/uc1.32106000156098.

139 ***the silica particles*** Joy Gardner, *Color and Crystals: A Journey Through the Chakras* (Freedom, CA: Crossing Press, 1988), 15.

140 ***the pure stone of love*** Robert Simmons, *The Pocket Book of Stones* (Rochester, Vermont: Destiny, 2015), 271.

140 ***calm and cleanse*** Simmons, *Pocket Book*, 271.

143 ***Magic's true purpose*** Bronislaw Malinowski, *Magic, Science and Religion* (Garden City, NY: Doubleday Anchor Books, 1954), reprint ed. (Mansfield Centre, CT: Martino Publishing, 2015), 90.

144 ***Robert Simmons, patron saint*** Robert Simmons, *Stones of the New Consciousness: Healing, Awakening, and Co-creating with Crystals, Minerals and Gems* (East Montpelier, VT: Heaven and Earth Publishing, 2009), 1.

144 ***partnered destiny*** Simmons, *New Consciousness*, 11.

144 ***W. T. Fernie's 1907*** W. T. Fernie, M.D., *Precious Stones: For Curative Wear, and Other Remedial Uses* (Bristol, GB: John Wright and Co, 1907), 18, https://hdl.handle.net/2027/hvd.32044023405038.

Object Permanence

150 ***Mourning a loved one*** Sigmund Freud, "Mourning and Melancholia" in *The Standard Edition of the Complete Psychological Works of Sigmund Freud*, trans. and ed. James Strachey, vol. 14 (London: Hogarth Press), 244.

150 ***This demand to renounce*** Freud, "Mourning and Melancholia," 245.

152 ***There the ego*** Freud, "Mourning and Melancholia," 257.

154 ***events that for a while are monotonous*** Carlo Rovelli, *The Order of Time*, trans. Erica Segre and Simon Carnell (New York: Riverhead Books, 2018), 103.

Camera Obscura

157 ***the empty chapel*** T. S. Eliot, "The Waste Land," Poetry Foundation website, accessed December 11, 2023, https://www.poetryfoundation.org/poems/47311/the-waste-land.

160 ***For methinks the understanding*** John Locke, *The Works of John Locke in Nine Volumes* (London: C. and J. Rivington, 1824), 142.

162 ***to have the right optical*** Carl Zimmer, "How the Eye Evolved," *New York Academy of Arts and Sciences* (Autumn 2009), accessed December 11, 2023, https://www.nyas.org/magazines/autumn-2009/how-the-eye-evolved.

162 ***One scientist called it*** Kate Hazlehurst and Lisa Henry, "Eyes on the Prize: The Evolution of Vision," Natural History Museum website, accessed December 11, 2023, https://www.nhm.ac.uk/discover/eyes-on-the-prize-evolution-of-vision.html.

162 ***And when our proverbial*** Malcom MacIver et al., "Massive Increase in Visual Range Preceded the Origin of Terrestrial Vertebrates," *PNAS*, January 24, 2017, accessed December 11, 2023, https://www.pnas.org/doi/10.1073/pnas.1615563114.

163 ***non-seeing armor*** Susan Mileus, "How to See with Eyes Made of Rock," *ScienceNews*, November 19, 2015, accessed December 11, 2023, https://www.sciencenews.org/article/how-see-eyes-made-rock.

163 ***a somewhat blurred*** Mileus, "How to See."

164 ***stain on the poor wife's cheek*** Nathaniel Hawthorne, "The Birthmark," in *Stories by Nathaniel Hawthorne*, ed. Arthur Ransome (London and Edinburgh: T. C. and E. C. Jack, 1908), 52.

165 ***ineludible gripe in which*** Hawthorne, "The Birthmark," 51.

165 ***the parting breath*** Hawthorne, "The Birthmark," 74.

167 ***when the task of gathering*** Howard Schwartz, "How the Ari Created a Myth and Transformed Judaism," Tikkun, March 28, 2011, accessed December 11, 2023, https://www.tikkun.org/how-the-ari-created-a-myth-and-transformed-judaism.

167 ***All Life, the Remnants*** Hans Jonas, "Immortality and the Modern Temper" in *The Phenomenon of Life: Toward a Philosophical Biology* (Chicago: University of Chicago Press, 1966), 273. For original psalm: "Let Us Worship the Spirit of the Paraclete," Manichaean Scriptures, The Gnostic Society Library, http://www.gnosis.org/library/Mani.html.

ACKNOWLEDGMENTS

When I first started writing these essays in 2017, after the death of my father, I wasn't sure what they would amount to or what form they would eventually take. I certainly didn't dare to think of myself as a writer of "personal essays." That these meditations have come together to form a collection at all seems like a small miracle, and it wouldn't have been possible without the many kind, patient, and insightful readers who helped me along the way.

Thank you to my agent, Rob McQuilkin, without whom I never would have become a writer at all, and whose commitment to this strange, niche collection has always been unwavering. In the middle of the journey of my life, you gave me an unlooked-for gift—the chance to be a writer—and it has made all the difference. Immense thanks to Ada Limón, Briallen Hopper, and David Ulin, who were kind enough to read early drafts of the collection and write blurbs in support of it. Your words of encouragement at this early phase of trying to put a book out into the world meant more to me than you can ever know.

Thank you to my colleagues in the Writing Program and the English Department at the University of Southern California, whose commitment to fight the good fight for literature every day in the classroom is an inspiration to me. A special thanks to David St. John for believing in me and inviting me into the English department fold. The opportunity to teach creative nonfiction to classes full of talented, eager young writers has helped me hone my craft in so many unexpected ways.

Thank you to the entire team at Milkweed Editions: Daniel Slager, for his faith in the writing; Lauren Langston Klein, for her timely and sensitive edits; Mary Austin Speaker, for her heart-stoppingly gorgeous cover design; and most especially to my editor, Helen Whybrow, who saw a glimmer of promise in my book proposal when she read it in 2021, and whose gentle guidance and support over the past three years have been essential in shaping the book.

Thank you to the magazine editors who published original versions of these essays, including Dana Snitzky at Longreads, Sam Dresser at *Aeon* magazine, Sarah Mesle at *Avidly*, and Bob Wilson at the *American Scholar*. Your readerly generosity and editorial wisdom have been invaluable supports along the way.

Thank you to the many friends who read versions of these essays as they evolved, including Tamara Griggs, Jessica Levenstein, and Mimi Long. Thank you to my friend Katie Mather, who built a website for her luddite friend and offered sensitive, generous readings of these essays in their various embryonic forms. Thank you to Julie Park, whose writing check-ins and pep talks kept me on task over the years, and whose own brilliant work in her book on camera obscuras helped me think about the nature of mind and interiority. Thank you to my sister Sarah, not only for her sharp readings of my essays, but for being a companion in navigating climate grief and a help in finding those places where the light breaks through. Thank you to my steadfast writing group, Kate Levin and Erika Nanes, who have provided the backbone of support for this book over the past five years. Against the odds of hectic schedules, heavy teaching loads, pandemics, and Los Angeles traffic, we somehow have always managed to carve out time for each other and our writing, and for this ongoing space of dialogue and friendship, I am immeasurably grateful.

Living in deep time tends to make one a poor attendant to the here and now. I am ever thankful to my dear husband, Jake, and my dear daughters, Sophia and Lydia, for extending me grace in this regard.

ELLEN WAYLAND-SMITH is the author of two books of American cultural history, *Oneida: From Free Love Utopia to the Well-Set Table* and *The Angel in the Marketplace: Adwoman Jean Wade Rindlaub and the Selling of America*. She lives in Los Angeles and is a professor of writing at the University of Southern California.

Founded as a nonprofit organization in 1980, Milkweed Editions is an independent publisher. Our mission is to identify, nurture, and publish transformative literature, and build an engaged community around it.

Milkweed Editions is based in Bdé Óta Othúŋwe (Minneapolis) within Mní Sota Makhóčhe, the traditional homeland of the Dakhóta people. Residing here since time immemorial, Dakhóta people still call Mní Sota Makhóčhe home, with four federally recognized Dakhóta nations and many more Dakhóta people residing in what is now the state of Minnesota. Due to continued legacies of colonization, genocide, and forced removal, generations of Dakhóta people remain disenfranchised from their traditional homeland. Presently, Mní Sota Makhóčhe has become a refuge and home for many Indigenous nations and peoples, including seven federally recognized Ojibwe nations. We humbly encourage our readers to reflect upon the historical legacies held in the lands they occupy.

milkweed.org

Milkweed Editions, an independent nonprofit literary publisher, gratefully acknowledges sustaining support from our board of directors, the McKnight Foundation, the National Endowment for the Arts, and many generous contributions from foundations, corporations, and thousands of individuals—our readers. This activity is made possible by the voters of Minnesota through a Minnesota State Arts Board Operating Support grant, thanks to a legislative appropriation from the arts and cultural heritage fund.

Interior design by Mary Austin Speaker
Typeset in Bely

Bely was designed by Roxane Gataud for the
TypeTogether foundry in 2014. Bely is designed
with classical proportions for maximum
legibility and received the Type Directors Club
Award of Excellence in Type Design in 2017.